Computer-Aided Glaucoma Diagnosis System

Computer-Aided Glaucoma Diagnosis System

Arwa Ahmed Gasm Elseid and Alnazier Osman
Mohammed Hamza

CRC Press
Taylor & Francis Group
Boca Raton London New York

CRC Press is an imprint of the
Taylor & Francis Group, an **informa** business

First edition published 2020
by CRC Press
6000 Broken Sound Parkway NW, Suite 300, Boca Raton, FL 33487-2742

and by CRC Press
2 Park Square, Milton Park, Abingdon, Oxon, OX14 4RN

© 2020 Taylor & Francis Group, LLC

CRC Press is an imprint of Taylor & Francis Group, LLC

Library of Congress Cataloging-in-Publication Data

Names: Elseid, Arwa Ahmed Gasm, author.
Title: Computer-aided glaucoma diagnosis system / Arwa Ahmed Gasm Elseid,
Alnazier Osman Mohammed Hamza.
Description: First edition. | Boca Raton, FL : CRC Press, 2020. | Includes
bibliographical references and index. | Summary: "Glaucoma is the second
leading cause of blindness globally. Early detection and treatment can
prevent its progression to avoid total blindness. This book discusses
and reviews current approaches for detection and examines new approaches
for diagnosis using CAD and machine learning techniques." -- Provided by publisher.
Identifiers: LCCN 2020003001 (print) | LCCN 2020003002 (ebook) | ISBN
9780367406264 (hardback) | ISBN 9780367406288 (ebook)
Subjects: LCSH: Glaucoma--Diagnosis--Computer simulation. | Diagnostic
imaging--Digital techniques. | Diagnostic imaging--Computer-aided
design.
Classification: LCC RE871 .E47 2020 (print) | LCC RE871 (ebook) | DDC
617.7/410754--dc23
LC record available at https://lccn.loc.gov/2020003001
LC ebook record available at https://lccn.loc.gov/2020003002

ISBN: 978-0-367-40626-4 (hbk)
ISBN: 978-0-367-40628-8 (ebk)

Typeset in Times LT Std
by Deanta Global Publishing Services, Chennai, India

Dedication

To my parents and family

Contents

Preface

Glaucoma is a group of eye diseases that cause optic nerve damage. Glaucoma has no symptoms, and if it is not detected at an early stage, it may cause permanent blindness. The disease's progression precedes some structural damage to the retina. Mainly, it is diagnosed by an examination of the size, structure, shape, and color of the optic disc (OD) and optic cup (OC), and the retinal nerve fiber layer (RNFL), which suffers from the subjectivity of human experience, fatigue factor, etc. The fundus camera is among one of the biomedical imaging techniques used to analyze the internal structure of the retina. With the widespread adoption of higher quality medical imaging techniques and data, there are increasing demands for medical image-based computer-aided diagnosis (CAD) systems for glaucoma detection, because human mistakes and other retinal diseases such as Age-related Macular Degeneration (AMD), can affect the early detection of glaucoma. Existing medical devices such as Optical Coherence Tomography (OCT) and Heidelberg Retinal Tomography (HRT) are very expensive for use in regular checkups. The proposed technique provides a novel algorithm to detect glaucoma from digital fundus images using a combined features set. An evaluation of the proposed algorithm is performed using a RIM_ONE (version two) database, containing fundus images from 158 patients (118 healthy and 40 glaucomatous images); Drishti_GS, which contains 101 fundus images (70 glaucomatous images and 31 healthy images); and RIM_ONE (version one), which contains 455 fundus images (200 glaucomatous and 255 healthy images) via MATLAB® software. The proposed system was used to detect glaucoma via 3 steps; firstly, OD and OC segmentation. In OD and OC segmentation, several steps were carried out such as pre-pressing, thresholding, boundary smoothing, and disc reconstruction to a full circle, where OD segmentation achieved a best Dice Coefficient (DSC) of 90% and a Structural Similarity (SSIM) of 83%, and OC segmentation results were a Dice Coefficient of 73% and a Structural Similarity (SSIM) of 93%. Secondly, shape, color, and texture features were extracted from segmented parts. The most relevant features were then selected. Finally, many classifiers were applied to find the best classification accuracy, which was the support vector machine (SVM). This research proposes a novel combination of color-based, shape-based, and texture features by extracting 13 shape features from disc and cup, extracting 25 texture features from RNFL (retinal nerve fiber layer) using the gray level co-occurrence method and Tamar algorithm, and 3 color features for each disc, cup, and RNFL. Next, best features were selected by t-test method and Sequential feature selection (SFS) to introduce 8 features with an average accuracy of 97%, maximize area under curve (AUC) 0.99, specificity 96.6%, and sensitivity 98.4% to the first database, and 91.5 and 94.5 to the second and third databases respectively, with a training time of 1.5623 sec and prediction time 2,600 obs/sec (one billionth of a second). The proposed algorithm performed excellently compared with previous studies (from 2011 to the present), both in features types and overall performance. The key contribution of this work is the proposed real-time algorithm for high accuracy glaucoma detection. The proposed method can make a valuable contribution to medical science by supporting medical image analysis for glaucoma detection. It would be beneficial for future works to design a complete, integrated, automated system to classify all of the different types of glaucoma, namely: Primary Open-Angle Glaucoma, Normal Tension Glaucoma, Angle Closure Glaucoma, Acute Glaucoma, Exfoliation Syndrome, and Trauma-Related Glaucoma, and to upgrade the system to compute the progress of the disease by comparing different images of the same patient to be used for follow up.

MATLAB® is a registered trademark of The MathWorks, Inc. For product information, please contact:

The MathWorks, Inc.
3 Apple Hill Drive
Natick, MA 01760-2098 USA
Tel: 508 647 7000
Fax: 508-647-7001
E-mail: info@mathworks.com
Web: www.mathworks.com

Acknowledgments

First and foremost, thanks should be expressed to Allah for his assistance and for giving me the opportunity to be engaged in this study, and the great help to overcome all the difficulties throughout this period of my research. I would like to thank my supervisor, Prof. Alnazier Osman Hamza, who helped me in completing this study with his great efforts, continued guidance, and assessment. Also I would like to express my deepest thanks to Dr. Fragoon Mohamed Ahmed for his precious advice and support. Special thanks goes to Dr. Ahmed Gasm Elseid and Dr. Egbal Suliman. Finally, I shall be eternally grateful to members of my family: my husband Eng. Sohaib Abd Algawy and my children. I thank all those who assisted and supported me during this study.

Authors

Dr. Arwa Ahmed Gasm Elseid is Assistant Professor in Biomedical Engineering, Khartoum, Sudan. Her research has centered on CAD (Computer-Aided Diagnosis) systems for glaucoma detection via digital fundus images, which developed a new algorithm that used image processing techniques (contrast equalization, filtration, edge detection and segmentation, feature extraction and selection), and then classified the analyzed data from the extracted features. A critical aspect of this work has led to the development of an accurate and inexpensive CAD system for glaucoma diagnosis using MATLAB® software.

Dr. Alnazier Osman Mohammed Hamza is Professor of Medical Imaging at the College of Engineering, Sudan University of Sciences and Technology, Khartoum, Sudan. He is married and has six children.

List of Abbreviations

ACG	Angle Closure Glaucoma
AMD	Age-related Macular Degeneration
AUC	Area Under Curve
BRIEF	Binary Robust Independent Elementary Features
CAD	Computer-Aided Diagnosis
CDR	Cup-to-Disc Ratio
CFI	Colors Funds Image
CSL	Cost-Sensitive Learning
CSLO	Confocal Scanning Laser Ophthalmoscopy
DOG	Difference of Gaussians
DSC	Disc Similarity Coefficient
DWT	Discrete Wavelet Transform
FA	Fluorescein Angiography
FAF	Funds Auto Fluorescein
FN	False Negative
FP	False Positive
FPR	False Positive Rate
GLCM	Gray Level Co-occurrence
GMRF	Gaussian Markov Random Fields
GRI	Glaucoma Risk Index
GUI	Graphical User Interface
HOS	Higher Order Spectra
HRT	Heidelberg Retinal Tomography
HSI	Hue, Saturation, and Intensity
IOP	Intraocular Pressure
ISNT	Inferior, Superior, Nasal, and Temporal
K-NN	K-Nearest Neighbors algorithm
LOG	Laplacian of Gaussian
MSE	Mean Squared Error
NTG	Normal Tension Glaucoma
OAG	Open Angle Glaucoma
OC	Optic Cup
OCT	Optical Coherence Tomography
OD	Optic Disc
ONH	Optic Nerve Head
PPA	Parapapillary Atrophy
PSNR	Peak Signal Noise Ratio
RBF	Radial Basis Function
RGB	Red, Green, Blue Channels
RNFL	Rental Nerve Fiber Layer
ROC	Receiver Operating Characteristic
SBFS	Sequential Backward Floating Selection

SBS	Sequential Backward Selection
SFFS	Sequential Forward Floating Selection
SFS	Sequential Forward Selection
SLIC	Simple Linear Iterative Clustering algorithm
SMOTE	Synthetic Minority Over-sampling Technique
SNR	Signal-to-Noise Ratio
SSIM	Structural Similarity
STD	Standard Deviation
SVM	Support Vector Machine
TN	True Negative
TP	True Positive
TPR	True Positive Rate
TT	Trace Transform
WHO	World Health Organization

1 Introduction

1.1 BACKGROUND OF THE CAD SYSTEM

Computer-aided diagnosis (CAD) has become one of the major research approaches in medical imaging science. In this book, the motivation and philosophy for the early development of CAD systems are presented together with the glaucoma CAD system. With CAD systems, radiologists and doctors use the computer output as a "second opinion," and then make the final decisions. CAD is a concept established by taking into account the importance of physicians and computer applications equally, whereas automated computer diagnosis depends on computer algorithms only (Doi *et al.*, 2017).

"Today, the role of computer-aided diagnosis is expanding beyond screening programs and toward applications in diagnosis, risk assessment and response to therapy." (Giger, 2010)

The computer-aided detection (CAD) or computer-aided diagnosis (CAD) is the computer software or application that helps doctors to take decisions swiftly (Doi, 2017), (Li and Nishikawa, 2015). Medical imaging contains data in images evaluated by doctors to analyze abnormality and diagnose diseases. Analysis of imaging in the medical field is a very important approach because imaging is the basic modality to diagnose diseases at the earliest opportunity, and image acquisition is almost always an external operation that does not cause harm. Imaging techniques like MRI, X-ray, endoscopy, ultrasound, etc., if acquired with high energy will provide a good-quality image, but they will harm the human body; however, images that are taken with less energy will sometimes be poor in quality and have a low contrast. CAD systems are used to improve the image quality, helping to detect the medical image's abnormality correctly and diagnose different diseases (Chen *et al.*, 2013). Computer-aided diagnosis (CAD) technology includes multiple science aspects, such as artificial intelligence (AI), computer vision, and medical image processing. The main application of CAD systems is in finding abnormalities in the human body and diagnosing diseases as a second opinion. For example, the detection of tumors is a typical application because if they are diagnosed early in basic screening, it will help prevent the spread of cancer (Giger, 2000).

1.2 OBJECTIVES AND SIGNIFICANCE OF THE CAD SYSTEM

The main goal of CAD systems is to identify signs of abnormality at the earliest possible opportunity where a human professional might fail to notice them. For example, in mammography, the identification of small lumps in dense tissue, finding architectural distortion, and prediction of mass type as benign or malignant by its shape, size, etc.

CAD is usually restricted to mark the visible abnormalities in images, whereas CAD helps to evaluate the abnormalities identified in the images at an earliest disease stage. For example, it highlights abnormalities in RNFL, breaking the ISNT rule, showing texture and color abnormalities in the retina as the earliest signs of glaucoma. This helps the radiologist to draw their conclusion. Though CAD has been used for over 40 years, it still does not reach its expected outcomes. We agree that CAD cannot act as a substitute for a real doctor, but it definitely helps radiologists become better decision makers. It plays as a second opinion and final interpretative role in medical diagnosis.

1.3   APPLICATIONS OF CAD SYSTEMS

Different CAD systems are used in the diagnosis of breast cancer, lung cancer, colon cancer, prostate cancer, bone metastases, coronary artery disease, congenital heart defect, pathological brain detection, Alzheimer's disease, glaucoma disease, and Diabetic Retinopathy.

1.4   PREVIOUS STUDIES OF CAD SYSTEMS

Halalli *et al.* (2017) proposed a CAD system to detect breast cancer where he addressed detection steps such as pre-processing, segmentation, feature extraction, and classification. He proposed a CAD used for identification of subtle signs for use in breast cancer detection and classification.

Hiroshi *et al.* (2008) proposed a CAD system for the early detection of (1) cerebrovascular diseases using brain MRI and MRA images by detecting lacunar infarcts, unruptured aneurysms, and arterial occlusions; (2) other types of ocular diseases such as glaucoma, diabetic retinopathy, and hypertensive retinopathy using retinal fundus images; and (3) breast cancers, using ultrasound 3-D volumetric whole breast data using the breast masses detection.

Na Young *et al.* (2014) proposed a study to evaluate the diagnostic performance of the CAD system in full-field digital mammography for detecting breast cancer when used by a dedicated breast radiologist (BR) and a radiology resident (RR), in order to detect who could most benefit from a CAD application. They found that CAD was helpful for the BRs to improve their diagnostic performance and for RRs to improve their sensitivity in a screening setting, and concluded that CAD could be essential for radiologists by decreasing reading time and improving diagnostic performance.

The CAD accuracy was generally better in the BR group than in the RR group, but sensitivity was better with CAD use in both groups, from 81.10% to 84.29% in the BR group, and from 75.38% to 77.95% in the RR group. The most improvement in disease diagnosis was observed in the BR group, whereas in the RR group sensitivity improved but specificity, PPV, and NPV did not. The main advantage of using the CAD system was shortened time in both the BR and RR groups, from 111.6 minutes to 94.3 minutes for BR, and 135.5 minutes to 109.8 minutes for RR, which was more significant for the RR group than it was for the BR group.

Godoy *et al.* (2013) proposed a study to evaluate the impact of CAD on the identification of sub solid and solid lung nodules on thin- and thick-section CT.

They used 46 chest CT examinations with ground-glass opacity (GGO) nodules; CAD marks computed using thin data were evaluated in two phases, and for 155 nodules they found (mean, 5.5 mm; range, 4.0–27.5 mm) – 74 solid nodules, 22 part-solid (part-solid nodules), and 59 GGO nodules – CAD stand-alone sensitivity was 80%, 95%, and 71%, respectively, with 3 false-positives on average (0–12) per CT study. Reader(thin)+CAD(thin) sensitivities were higher than reader(thin) for solid nodules (82% vs. 57%, $p < 0.001$), part-solid nodules (97% vs. 81%, $p = 0.0027$), and GGO nodules (82% vs. 69%, $p < 0.001$) for all readers ($p < 0.001$). Respective sensitivities for reader(thick), reader(thick)+CAD(thick), reader(thick)+CAD(thin) were 40%, 58% ($p < 0.001$), and 77% ($p < 0.001$) for solid nodules; 72%, 73% ($p = 0.322$), and 94% ($p < 0.001$) for part-solid nodules; and 53%, 58% ($p = 0.008$), and 79% ($p < 0.001$) for GGO nodules. For reader(thin), false-positives increased from 0.64 per case to 0.90 with CAD(thin) ($p < 0.001$) but not for reader(thick); false-positive rates were 1.17, 1.19, and 1.26 per case for reader(thick), reader(thick)+CAD(thick), and reader(thick)+CAD(thin), respectively.

Automating mass chest screening for tuberculosis (TB) requires segmentation and texture analysis in chest radiograph images (Bram, 2001). Several rule-based schemes, pixel classifiers, and active shape model techniques for segmenting lung fields in chest radiographs are described and compared. An improved version of the active shape model segmentation technique, originally developed by Cootes and Taylor from Manchester University, UK, is described that uses optimal local features to steer the segmentation process and outperforms the original method in segmentation tasks for several types of medical images, including chest radiographs and slices from MRI brain data. In order to segment the posterior ribs in PA chest radiographs, a statistical model of the complete rib cage is constructed using principal components analysis and a method is described to fit this model to input images automatically. For texture analysis, an extension is proposed to the framework of locally orderless images, a multi-scale description of local histograms recently proposed by Koenderink and Van Doorn from Utrecht University, the Netherlands. The segmentation and texture analysis techniques are combined into a single method that automatically detects textural abnormalities in chest radiographs and estimates the probability that an image contains abnormalities. The method was evaluated on two databases. On a 200-case database of clinical chest films with interstitial disease from the University of Chicago, excellent results were obtained in the area under the ROC Curve 50.99. The results for a 600-case database from a TB screening program are an encouraging area under the ROC Curve 50.82.

The aim of the Marco *et al.* (2004) study was to evaluate diagnostic accuracy provided by different statistical classifiers on a large set of pigmented skin lesions grabbed by four digital analyzers located in two different dermatological units. Experimental Design: images of 391 melanomas and 449 melanocytic nevi were included in the study. The methodology was based on a linear classifier to identify a threshold value for a sensitivity of 95%, therefore, a K-nearest-neighbor classifier, a nonparametric method of pattern recognition, was constructed using all available image features and trained for a sensitivity of 98% on a large dataset of lesions. The obtained result on independent test sets of lesions for the linear classifier and the K-nearest-neighbor classifier produced a mean sensitivity of 95% and 98%, and a mean specificity of 78% and of 79%, respectively. They then suggested that computer-aided differentiation of melanoma from benign pigmented lesions obtained with DB-Mips is feasible and, above all, reliable. In fact, the same instrumentations used in different units provided similar diagnostic accuracy. Whether this would improve early diagnosis of melanoma and/or reduce unnecessary surgery needs to be demonstrated by a randomized clinical trial.

1.5   HOW TO EVALUATE A CAD SYSTEM

The evaluation of a CAD system is very important and can be carried out in several ways, such as an analysis of data generated by the CAD system in a laboratory, or by comparing the system output with the radiologist performance in an actual clinical practice. There are many aspects to evaluate, such as:

1.5.1   SENSITIVITY AND SPECIFICITY

The sensitivity and specificity can be measured by observing the output of a CAD system on a set of "truth" cases. Truth is generally established by the presence (e.g., cancer) or absence (e.g., clinical follow-up) of the disease. The sensitivity is determined based on the percentage of positive cases in which the CAD system diagnoses it as positive, but specificity is determined by the number of false CAD marks per normal image or case. These evaluation criteria are dependent on the case collection, bias (whether it is intended or not), and sometimes the diagnosing results vary from one CAD to another based on the type of input data. Thus, the same CAD algorithm can demonstrate varying sensitivities and specificities depending on the input data or images. For that the best method to compare CAD systems is to determine the sensitivity and false marker rates on the same set of "truth" cases. These cases must be at the test step for the CAD system, that is, they should not have been used to train the CAD algorithms. The number of cases will be important in order to establish the statistical significance of superiority or equivalence in performance when comparing CAD systems.

1.5.2   LABORATORY STUDIES

This type of evaluation depends on radiologists or other health care professionals to evaluate a set of "truth" cases to determine the sensitivity of the CAD and compare the results before and after the CAD system with the professionals diagnosing. Such considerations are valuable to survey the potential advantage of the CAD and give appraisals of the CAD system detection rate and workup/review rates.

1.5.3   ACTUAL CLINICAL PRACTICE EXPERIENCE

From numerous points of view, these outcomes may be viewed as the best appraisal of a CAD framework, in that they assess the commitment (or scarcity in that department) of CAD in a genuine clinical work setting. In this circumstance, the effect of

CAD on the recognition of diseases and its accuracy rate are evaluated. This information will be most helpful in system evaluation compared with similar clinical research, and the total reports from various practices help to show a pattern on the worth of the presented CAD system. This should be possible in a "successive read" clinical preliminary, in which the test is first perused preceding, and then afterward following, CAD input [2–5]. The CAD systems evaluation can be based on the system accuracy, minimize the workload, and the disease diagnosing.

1.6 BACKGROUND OF THE GLAUCOMA DISEASE

Glaucoma is dangerous as an ocular disease because it is the second-leading cause of blindness with about 60 million glaucomatous cases globally (Lim *et al.*, 2010), and it is responsible for 5.2 million cases of blindness based on (Lim *et al.*, 2012), with more than 90% of the patients unaware of the condition (Zhang *et al.*, 2014). In 2014, the World Health Organization (WHO) reported that 285 million people were estimated to be visually impaired worldwide: 39 million were blind and 246 million had low vision, in which 80% of all visual impairment could be prevented or cured. The WHO also stated that around 90% of the world's visually impaired people lived in low-income settings (Koprowski, 2014). Clinically, glaucoma is a chronic eye disease that damages the optic nerve progressively as the disease progresses, causing more optic head damage due to loss of peripheral vision and resulting in a gradual loss of vision. Finally, glaucoma is associated with total blindness. Glaucoma must be managed at the early stage to prevent irreversible optic nerve damage. Treatment can prevent progression of the disease. Therefore, early detection of glaucoma is important to prevent blindness.

Currently, there are three methods for detecting glaucoma: assessment of abnormal visual field, assessment of intraocular pressure (IOP), and assessment of optic nerve damage. The visual field check requires special equipment that is usually present only in hospitals. It is a subjective examination which assumes that patients fully understand the testing instructions, and that they are able to cooperate and complete the test. Moreover, the information obtained may not be reliable for children or ICU patients.

In the second method, a large proportion of glaucoma patients have a normal level of IOP. Thus, IOP measurement is neither specific nor sensitive enough to be used for the effective screening for early glaucoma. The assessment of optic nerve damage is the best of the other two methods (Zhang *et al.*, 2012). The main methods for optic nerve assessment by trained specialists are 3-D imaging techniques such as Heidelberg Retinal Tomography (HRT) and Ocular Computing Tomography (OCT). However, optic nerve assessment by specialists is subjective and the availability of HRT and OCT equipment is limited due to its expense. In summary, there is still no systematic and economic way of detecting glaucoma. There is a need for an automatic and economically viable system for the detection of glaucoma in an accurate way, possibly by using the digital color fundus image in Figure 1.1, which is a more cost-effective imaging modality to assess optic nerve damage compared to HRT and OCT, and has been widely used in recent years to diagnose various ocular diseases including glaucoma. In this research, a system for diagnosing glaucoma from non-glaucoma cases via fundus images analysis has been proposed.

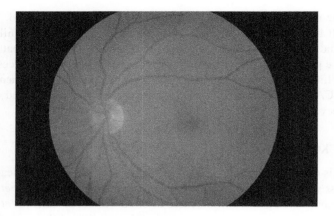

FIGURE 1.1   Shows an example of fundus image obtained from a fundus camera used for glaucoma detection (www.Shutterstock.com).

Glaucoma is one of the most common causes of permanent blindness worldwide. Nonetheless, if glaucoma is diagnosed early enough, it can be properly managed to prevent a major loss of vision; although there is no cure for glaucoma, medication can be used to prevent vision loss (Lim *et al.*, 2012). Following are six tests used to help detect glaucoma mentioned at www.biomedical-engineering-online.com.

1.6.1 Tonometry

Tonometry tests are used to measure the pressure inside the eye, also known as intraocular pressure (IOP). Having eye pressure that is higher than normal counts as a risk factor for glaucoma, but it does not mean a definite diagnosis of glaucoma.

1.6.2 Ophthalmoscopy

Ophthalmoscopy is a device used to examine the inside part of the eye and can be performed on a dilated or non-dilated eye. The color, shape, and overall health of the optic nerve are important signs in glaucoma diagnosis that can be checked directly or by using a digital camera to photograph the optic nerve.

1.6.3 Gonioscopy

Gonioscopy is a test for examining the angle where the cornea meets the iris, using a special mirrored device to gently touch the surface of the eye. Whether this angle is open or closed can show the doctor what type of glaucoma there is and how severe it may be.

1.6.4 Visual Field Testing

Visual field testing (perimetry) is a test that measures the patient's vision. The patient will look straight ahead at a small light and will then be asked to tell the examiner when he sees a light flash to the side of his peripheral vision.

1.6.5 NERVE FIBER ANALYSIS

Nerve fiber analysis is a test for glaucoma in which the thickness of the nerve fiber layer is measured. Thinner areas may indicate damage caused by glaucoma, especially for patients who are suspected of having glaucoma, and can be used to indicate if the glaucoma is worsening.

1.6.6 PACHYMETRY

Pachymetry is used to measure the thickness of the cornea because it reflects the influence on the eye pressure. But research is still being conducted on the importance of corneal thickness in glaucoma testing.

1.7 COMPUTER-AIDED DIAGNOSIS IN GLAUCOMA

Computer-aided diagnosis (CAD) is the way to automate the detection process for glaucoma disease, which has attracted extensive attention from clinicians and researchers. It not only alleviates the burden on the clinicians by providing an objective opinion with valuable insights, but also offers early detection and easy access for patients. By reviewing ocular CAD methodologies for various data types for each data type, and investigating the databases and algorithms to detect different ocular diseases, the advantages and shortcomings has been found, and analyzed, to find three types of data (i.e., clinical, genetic, and imaging) that have been commonly used in existing methods for CAD.

The recent developments in methods used in CAD for ocular diseases (such as diabetic retinopathy, glaucoma, age-related macular degeneration, and pathological myopia) are investigated and summarized, and the CAD for glaucoma disease has shown much progress over the past years (Zhang and Khow, 2012).

There have been surveys on retinal imaging in the area of ocular research (Abràmoff et al., 2010), (Bernardes et al., 2011). However, there is still a shortage of CAD systems for ocular disease diagnosis. This has motivated us to write a systemic method for CAD in glaucoma disease.

1.8 PROBLEMS OF THE STUDY

In the cases of glaucoma in which the visual field test result and IOP assessment are not reliable, e.g., for advanced AMD disease (Age-related Macular Degeneration), the visual field measurement is usually carried out using a static perimeter, which is difficult for different diseases and it needs patient cooperation to follow the instructions. In this case, the IOP is not sensitive because in some kinds of glaucoma the pressure is normal, which is called normal tension glaucoma, and the best ways to detect glaucoma via ONH and imaging modality (OCT, HRT) are expensive. It is then necessary to search for other methods which will enable a determination of the glaucoma disease. One such method is the analysis of digital fundus images of the eye fundus taken via a fundus camera device.

Where studies indicate that radiologists do not detect all abnormalities on images that are visible on retrospective review, they do not always correctly characterize abnormalities that are found. In the clinical interpretation of medical images, limitations in the human eye–brain visual system, reader fatigue, distraction, the presence of overlapping structures that camouflage disease in images, and the vast number of normal cases seen in screening programs provide cause for detection and interpretation errors (Adrian *et al.*, 2012). For the previously mentioned reasons, such as human mistakes, other diseases affect glaucoma diagnosis and founded devices are expensive (OCT, HRT) there is a need for CAD system to diagnose glaucoma, mainly on the basis of optic nerve head (ONH) and retinal nerve fiber layer (RNFL) damage in the digital fundus image. Zhou *et al.* (2012), who investigated the use of MRMR-based Feature Selection in automatic glaucoma diagnosis, used data from heterogeneous data sources, i.e., retinal image and eye screening data to facilitate a better understanding of the CAD system and improve data collection for glaucoma research.

1.9 OBJECTIVES

In this research, a review of the current approaches for automatic glaucoma detection is presented, and new features for detecting glaucoma in fundus images are proposed, and the recommended methods of diagnosis, challenges, and the existing automated methods will be surveyed. Based on recent successful machine learning techniques and ophthalmologists' observations, the new learning-based techniques for glaucoma detection are investigated and proposed. The RIM-ONE database will be used to evaluate the accuracy of the proposed methods.

1.9.1 GENERAL AIM

The general aim of this project is to develop a reliable automated computer-aided diagnosis system (CAD) for detecting glaucoma, and develop an algorithm that detects the optic disc and optic cup.

1.9.2 SPECIFIC OBJECTIVES

1. Choose the best subset of features that describes the optic nerve and detect the glaucoma disease.
2. Assess the accuracy, sensitivity, and specificity of the classifier.

1.10 RESEARCH METHODOLOGY

This thesis presents a detection strategy motivated by the evaluation guidelines used by ophthalmologists for the diagnosis of glaucoma. The proposed strategy can be seen as an approach for integrating multiple features gathered from the analysis of optic disc, cup, and RNFL indicators from retinal fundus images. The analysis part consists of quantification of changes that happened in OD, OC, and RNFL.

The resultant parameters of the analysis components encode information to detect the glaucoma disease from the obtained features, along with the differences between glaucoma and healthy images. The final decision on the presence of glaucoma is made by the classification of these features and evaluating the results compared with the ground truth. Figure 1.2 illustrates the proposed detection approach for detecting glaucoma in digital fundus images based on employing many of the publicly available image datasets with the diagnosing and segmentation of ground truth. The chosen datasets act as a base in order to provide robust and consistent evaluations.

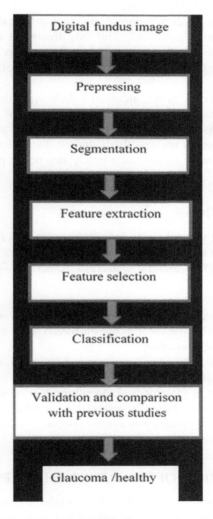

FIGURE 1.2 Shows the proposed algorithm approach for detecting glaucoma in digital fundus images.

1.11 SIGNIFICANCE OF THE STUDY

- This study provides an expert system for real-time fundus image analysis, which gives radiologists an opportunity to improve their image interpretation. The proposed system will improve the diagnosis of glaucoma and will therefore minimize the miss detection rate and help in early diagnosis and treatment, which can significantly improve the chance of managing the glaucoma disease. The results and images of the diagnosis can be stored in a digital format (an image file for the retina photo and a medical report) and used to generate a diagnosis database useful for researchers, for medical practice, and for patient follow-up.
- A database with patients' personal and care information can be developed with a secure password. The long treatment costs, supplementary tests, and founded images will help the doctors in follow-up with the patient. The quality and accessibility of eye medical services will increase.
- Finally, an accurate, novel, automated technique for glaucoma detection will be designed that reduces the workload of ophthalmologists, checks the disease's progression, is easy to operate, and is inexpensive. Thus, once glaucoma is correctly diagnosed, the patient can take proper medicine or undergo surgery in a timely manner to prevent permanent blindness.

1.12 CONTRIBUTIONS

The contributions for this book come from the novel digital fundus image analysis solutions developed for different glaucoma indicators, and are summarized below.

An automatic CAD system for glaucoma detection with high accuracy is developed:

1. The optic disc segmentation algorithm is modified by a circle reconstruction approach, and is used to improve the thresholding model and extended to the optic cup segmentation.
2. An algorithm is developed to detect cup and disc color, shape features, and RNFL texture features.
3. The performance of the proposed algorithms is presented. Cup, disc, and RNFL features are evaluated on several databases for glaucoma diagnosis. A more powerful algorithm may further improve classification accuracy.
4. A new features set will be used to describe the disease.

1.13 ORGANIZATION OF THE BOOK'S CHAPTERS

The outline of the book's thesis is in Chapter 2 a brief review of the medical background is presented, followed by the theoretical and mathematical background used in fundus image processing. Chapter 3 features a literature review about segmentation and feature extraction. Chapter 4 contains the formulation of the proposed methodology. In Chapter 5 the results of the optic disc and optic cup segmentation algorithm are presented, and the feature extraction and selection method's experimental results and performance evaluations of the classifier are given. Chapter 6 contains the conclusions and a discussion.

2 Medical and Mathematical Background

2.1 THE HUMAN EYE

In humans, the eye is a specialized sense organ capable of receiving visual images, which are then carried to the brain, see Figure 2.1 (www.shutterstock.com). This provides the ability to see color, detect motion, identify shapes, gauge distance and speed, judge the size of faraway objects, and see them in three dimensions, even though images fall into two dimensions (Ali Allam, 2017).

2.1.1 STRUCTURE OF THE HUMAN EYE

The human eye, as seen in Figure 2.1, is a complex structure, where light passes through the cornea and a transparent crystalline lens, which assists in the focusing of light onto the retina through the pupil.

The amount of light that enters the eye is controlled by the iris, which has the ability to enlarge or contract, while the lens converges the incoming light rays to a sharp focusing point onto the retina by lengthening and shortening its width. The retina, situated at the back of the eye, is approximately 0.5 mm thick, see Figure 2.2. In the center of the retina is the optic nerve, an oval white area measuring about 2 mm (height) × 1.5 mm (width). At approximately (4.5–5 mm), or 2.5 disc diameters away from the optic nerve, lies a blood vessel-free reddish region, known as the fovea, Figure 2.2. The center of the fovea is also called the macula.

The retina is a multi-layered sensory tissue that lies at the back of the eye, and contains millions of photoreceptors that capture light rays and convert them into electrical impulses that will in turn convert them into images in the brain. The main photoreceptors in the retina are rods and cones. Rod cells are very sensitive to changes in contrast, even at low light levels, hence they are able to detect movement, but they are imprecise and insensitive to color located in the periphery of the retina and are used for scotopic vision (night vision). Cones are high precision cells capable of detecting the colors that are concentrated in the macula, the area responsible for photopic vision (day vision). The very central portion of the macula is called the fovea, which is where the human eye is able to best distinguish visual details. The loss of peripheral vision may cause damage to the macula and can result in the loss of central vision.

The optic nerve on the retina is also called the blind spot, as it is insensitive to light. Often, the fovea, optic nerve, retina, and retinal vasculature are referred to as the ocular fundus structures.

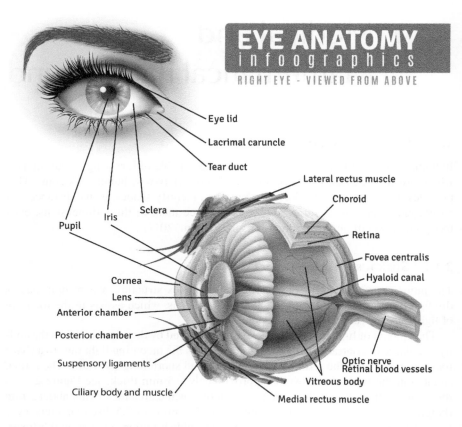

FIGURE 2.1 Shows the human eye's inside and outside structures: www.shutterstock.com

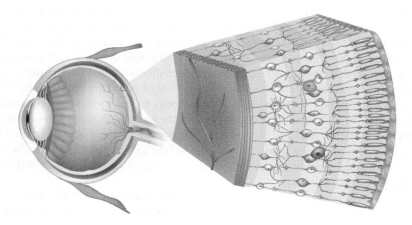

FIGURE 2.2 Shows the anatomy and schematic diagram of the human eye – Sagittal view of the human eye with a schematic enlargement of the retina, and schema of the layers of the developing retina (www.shutterstock.com).

2.2   IMAGE ACQUISITION OF THE RETINA

2.2.1   Fundus Image Capture

A color digital retinal fundus camera is widely used to capture an image of the retina. The fundus camera is a widely used and complex optical imaging device with a low-power microscope and an attached camera. The general working principle described below is based on the overview provided in Tyler *et al.* (2014).

The interior surface of the eye (i.e., retina, vasculature, optic disc, and macula) is captured using a specialized low-power microscope with an attached camera. Moreover, in stereo fundus photography, image intensities represent the amount of reflected light from two or more different viewing angles for depth resolution.

Light is first generated from either a viewing lamp or electronic ash, before being passed through a set of filters, mirrors, and a series of lenses for focusing. A mask on the uppermost lens is then used to shape the light into a doughnut. Based on the Gullstrand principle, the ring of light is projected on the cornea, through the pupil. The resulting retinal image then leaves the cornea through the unilluminated portion of the doughnut. This space within the ring allows a separation of both the incoming and outgoing illumination. The outgoing light continues through the central aperture of the mirror, through the astigmatic correction device and the diopter compensation lenses, and then back to the single lens reflex camera system.

Clinically, this camera is used by ophthalmologists and trained medical professionals to monitor and discover evidence of ocular abnormalities for immediate feedback, diagnosis, and treatment of retinal diseases. The fundus photographs are then kept as visual records to document the ophthalmoscope appearance of a patient's retina.

In the market, fundus cameras are described by their angle of coverage. This is derived from the optical angle of acceptance of the fundus camera lens and can range between 20° to 140°. A 30° is considered to be the normal angle of view, and creates a film image with a magnification of 2.5×, but wide-angle fundus cameras capture images between 45° and 140° with less retinal magnification. Capturing the retinal fundus photographs can be with a dilated or non-dilated pupil. The main advantage of dilation is to allow for a better view and image capture of the retina. However, even with dilation, the quality of the fundus image can still be affected by additional difficulties, such as the media opacity due to cataracts. Non-mydriatic retinal fundus cameras allow for digital photographs of the eye to be captured through a small pupil size (between 2.0 and 4.0 mm) without the need for the discomfort caused by pupil dilation.

2.2.2   Other Imaging Devices

  i. Fluorescein Angiography (FA); Indo-cyanine Green Angiography (ICG): shown in Figure 2.3, a grayscale image is produced where the blood flow within both the retina and the choroid is captured by injecting fluorescein dye and indo-cyanine dye, respectively, into the blood vein.
  ii. Fundus Auto Fluorescence (FAF): shown in Figure 2.4, the retina is illuminated with blue light which causes certain cellular components to glow without injecting any dye in the blood veins.

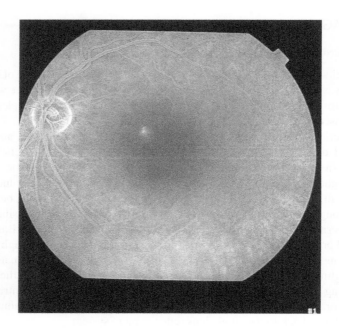

FIGURE 2.3 Images produced via angiography as a type from the fundus image (www. shutterstock.com).

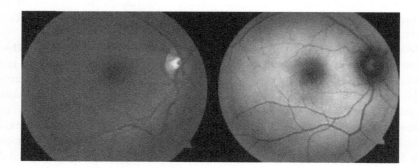

FIGURE 2.4 Images produced via Fundus Auto Fluorescence device as a type from the fundus image (Booysen, 2013).

   iii. Confocal Scanning Laser Ophthalmoscopy (CSLO): commercially, this imaging modality is known as Heidelberg Retina Tomography (HRT). It uses a special laser beam that is focused on the surface of the optic nerve in order to precisely capture a 3D image of the optic disc and the surrounding retina. HRT is a powerful diagnostic tool, particularly for glaucoma.

   iv. Optical Coherence Tomography (OCT): shown in Figure 2.5. Technically, OCT is not literally considered to be a "fundus imaging" modality as it is analogous to ultrasound, except for the fact that it utilizes light instead of

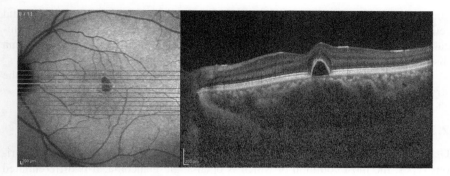

FIGURE 2.5   Images produced via Optical Coherence Tomography, which can be used to detect glaucoma (www.shutterstock.com).

sound. An OCT scan is used to capture the thickness of the retinal tissue by measuring the flight time of the originated backscatter. Thereby, it is well suited to monitor pathological conditions, such as macular edema, which leads to the swelling or thickening of the macula; a summarized comparison table of retinal acquisition devices is shown in Table 2.1.

In addition, OCT machines are unable to provide progression analysis as there is no means to measure the same tissue during follow-up exams. On the other hand, digital fundus images provide the advantages of full color, which helps to distinguish between cupping and pallor, and has a stable technology and widespread usage for

TABLE 2.1

Overview Comparison of Retinal Acquisition Devices, Used to Capture the Retina and Detect Glaucoma to Illustrate the Reason for Choosing the Digital Fundus Camera in This Research

Devices	Advantages	Limitations
Fundus camera	Stable technology. True color of retina. Low cost and widespread use.	Does not capture RNFL layer thickness or information.
GDx nerve fiber analyzer	Provide thickness of nerve fiber layer, which is important in glaucoma diagnosis.	Affected by existing eye conditions.
Heidelberg Retinal Tomography (HRT)	Provide topographic (layer thickness) image of optic nerve.	Topographic image is an approximate representation.
Optical Coherence Tomography(OCT)	Multiple cross-sectional (A-scan) of retina. Cross-sectional Tomograph (B-scan). Sub-layer retinal tissue thickness, which is important in glaucoma diagnosis.	Results change with new generations of the device and are not compatible. High cost.

community-level screening (Cello *et al.*, 2000). Furthermore, the record of a digital retinal photograph will not go out of date and is often used as a baseline for clinical evaluation and comparison. This makes it the ideal media to use for glaucoma disease screening purposes.

This section will cover an overview of the medical aspects of glaucoma and the structural changes it brings to the retina as it progresses (Bourne *et al.*, 2016).

2.3   THE GLAUCOMA DISEASE

Glaucoma is a collection of optic neuropathies. It is a chronic disease that has various types, the most common of which is open-angle glaucoma. It is differentiated from ACG by the appearance of the iridocorneal angle. In the OAG, the iridocorneal angle is open and has a normal form. On the other hand, the iridocorneal angle is closed in ACG. Glaucoma is further divided into primary and secondary. Primary glaucoma is characterized by the absence of additional ocular/systemic impairments. Regardless of the common features between primary and secondary glaucoma, however, secondary glaucoma might proceed differently. In addition, secondary glaucoma is accompanied by ocular/systemic diseases which could lead to the initiation of glaucoma.

The number of glaucoma patients – based on OAG and ACG, the two most widely spread types of glaucoma – totaled 60.5 million people in 2010. OAG percentage was 74% (44.7), while ACG cases made up the remaining 26% (15.7).

The number of people with bilateral blindness is shown for the total and the individual glaucoma types. The glaucomatous bilateral blindness is estimated to be 8.4 million (with 4.5 million and 3.9 million people affected by OAG and ACG, respectively). The epidemiology of glaucoma is expected to rise significantly in 2020 to 79.6 million people among the 11.2 million blind people. OAG contributes to 5.9 million cases of vision loss (from 58.6 million) and the ACG share is increased to 5.3 cases of vision loss (from 21 million). The glaucoma distribution numbers are based on the work by Quigley (1999). The various entities of glaucoma have common characteristics that are used to identify the presence of the disease. These include defects in the visual field, excavation of the optic disc, and optic nerve degeneration. IOP is a highly relevant feature of glaucoma. However, 50% of people diagnosed with OAG do not have ocular hypertension (Sommer *et al.*, 1991; Deepikaa *et al.*, 2016).

Moreover, another type of OAG is the NTG, where the IOP of the patients is always in the normal range. Nevertheless, reducing the IOP level has shown to delay or stop the progression of glaucoma even in NTG. Despite the attempts aiming to provide a precise definition of glaucoma like the case definition in Foster *et al.* (2002), the mechanisms of glaucoma are not completely understood, and there are still many challenges ahead for the scientific community to understand the pathology of glaucoma. Furthermore, it is argued that, due to the complex nature of glaucoma, all the aforementioned characteristics do not provide a clear identification for glaucoma, and patients could still be wrongly diagnosed with glaucoma or glaucoma could go undetected (Kroese and Burton, 2003).

2.3.1 GLAUCOMA AND THE VISUAL PATHWAY

The functional and morphological disorders due to glaucoma at the eye level have been extensively studied. However, the pathogenesis of glaucoma indicates the potential of extending the impairment to the rest of the visual system (Gupta and Yucel, 2007); for example, spreading atrophy along the visual pathway. This includes the intracranial optic nerve, lateral geniculate nucleus, and visual cortex (Gupta *et al.*, 2006). The rapid development of neuroimaging techniques during the last decades allowed the identification of the human visual system in-vivo and non-invasively. This was utilized in recent studies examining glaucoma. The optic radiation was examined in glaucoma patients and the neuronal density was reported to decrease when using size attenuation compared to normal subjects (Engelhorn *et al.*, 2011). A study produced glaucoma artificially in rats and found a correlation between certain parameters indicating the cerebral optic nerve fibers (Huajun *et al.*, 2007). Garaci *et al.* (2009) evaluated the integrity of the white matter fibers and axonal structure of the optic nerve, as well as the optic radiation in the presence of glaucoma. The fibers were compromised, and the degree of degeneration in the optic nerve was found to be in correlation with the severity of the glaucoma (Garaci *et al.*, 2009).

2.4 GLAUCOMA DIAGNOSIS

The clinical examination of glaucoma has a wide variety of modalities that contribute to the identification of the disease. This variation of modalities arises from the complex nature of the glaucoma pathology, where no single modality can provide a definite decision. For example, measuring the IOP as a major risk factor for glaucoma is not sufficient because its increase could be due to other diseases. The examination relies on evaluating the major features of glaucoma, which are the visual function and the appearance of the optic disc. In addition, ocular hypertension is an important indicator of the likelihood of having glaucoma and a determining factor for its progression path.

2.4.1 OPTIC NERVE HEAD

Fundus cameras take photographs of the interior surface of the eye, detailing the vessel tree and the optic disc among other structures. Digital fundus image scans can be used to detect the excavation of the optic disc and the reduction of the rim area, which are both significant signs of glaucoma. One of the most important parameters for glaucoma diagnosis is the cup-to-disc ratio. Two example fundus images of normal and glaucomatous subjects are shown in Figure 2.6. The rim thinning can be observed for the glaucomatous case. Despite the fact that fundus images are two-dimensional images, acquiring stereo images can provide three-dimensional information. The Heidelberg Retina Tomograph (Heidelberg Engineering, Heidelberg, Germany) utilizes the principals of confocal scanning laser ophthalmic copy to provide information about the topography of the retina surface. The topography is obtained by imaging sections of the retina, which are used to reconstruct a three-dimensional

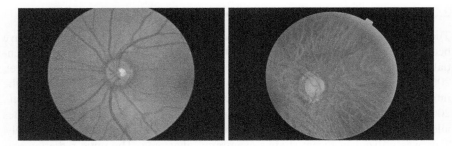

FIGURE 2.6 Fundus images showing the background of the retina for a healthy subject (a) and a glaucoma (b) patient. (www.shutterstock.com).

view of the ONH. This allows for a better representation and quantification of the optic disc. Glaucoma-relevant variables are extracted using HRT, such as horizontal and vertical cup-to-disc ratio, volume of cup and rim, and average RNFL thickness. The HRT parameters were shown to be sensitive for glaucoma diagnosis (Ferreras *et al.*, 2008). The HRT acquisitions of normal and glaucomatous subjects are demonstrated. Glaucomatous signs can be observed in the glaucoma patient. Optical coherence tomography relies on measuring an interference pattern from following a reference light and a light reflected from the retina. The light used can penetrate the retinal layers, providing depth information. The reflected light depends on the tissue structure and, thus, the components of the retina can be separated using OCT. Two- and three-dimensional images of the retina are obtained by combining depth scans. The thickness of the RNFL demonstrated a high ability to screen glaucoma in its early stages (Bowd *et al.*, 2001; Nouri and Mahdavi, 2004).

2.4.2 INTRAOCULAR PRESSURE

The instrument used for measuring the IOP is called the tonometer. Different techniques are utilized for tonometry. The main idea behind the application of tonometry, the most common type of tonometry, is to directly apply a force to flatten a region on the cornea. The force required is related to the ocular pressure. This method is relatively accurate and widely integrated in the clinical flow. A less precise variation of this technique is the non-contact tonometry which is usually used for screening purposes. In this procedure, the corneal curvature is reduced by the application of an air pulse and the force is similarly measured. In addition to identifying the risk of ocular hypertension, tonometry can be used to evaluate the treatment of glaucoma and its effect on the IOP.

2.4.3 VISUAL FIELD FUNCTION

The glaucomatous vision defects are located on the periphery of the visual field in its early stages. Perimetry is a widely used technique to examine the visual field. It detects the sensitivity of the eyes to identify light spots on a background at various positions in the visual field. It is used to screen early glaucoma patients and to diagnose moderate and advanced glaucoma by capturing the functional loss in vision (Cello *et al.*, 2000). From their research, Sample *et al.* suggested the incorporation of more than one functional test to enhance the diagnosis of glaucoma (Sample *et al.*, 2000). Moreover, they pointed out that the structural ONH damage, as shown in Figures 2.7 and 2.8, and the functional impairment due to glaucoma, have no definite

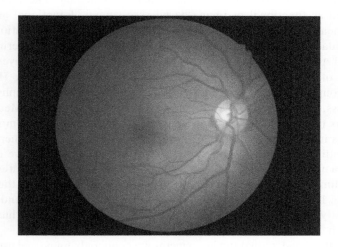

FIGURE 2.7   Anatomy of the optic nerve head – clinical features of the optic nerve (www. shutterstock.com).

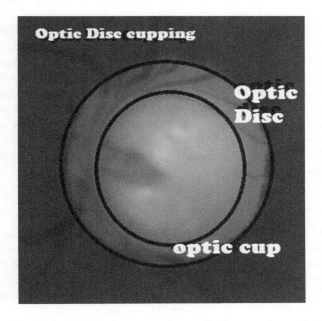

FIGURE 2.8   Progression of OC enlargement changes in glaucoma (www.shutterstock.com).

precedence; i.e., in some cases, visual function loss could be detected before optic disc abnormalities. And, in other cases, this sequence is reversed.

2.4.4 GLAUCOMA ASSESSMENT

Eye doctors usually use several tests to detect glaucoma. These are usually categorized as functional or image-based assessments. Functional assessment includes measurement of IOP (tonometry) and visual field examination. Image-based evaluation relies on optic nerve head imaging devices to determine optic nerve head structural damage or thinning of the retinal nerve fiber layer (RNFL). Examples of such imaging devices are Retinal Fundus Camera, Heidelberg Retinal Tomography (HRT), and Optical Coherence Tomography (OCT). Nonetheless, most of these methods have their own limitations. Measurement of IOP was reported to have a poor sensitivity of around 50% (Sommer *et al.*, 1991). This is partially due to cases of normal tension glaucoma, where patients have a condition in which optic nerve damage and vision loss have developed even with a normal pressure inside the eye. The visual field examination is often time-consuming and found to be unreliable due to factors such as the patient's fatigue and learning effects. Image-based evaluation using HRT and OCT are highly costly and are usually only available at tertiary hospitals, thus limiting its outreach.

On the other hand, retinal fundus cameras are commonly found and used primarily for the assessment of the retinal and optic nerve. Unlike IOP, the appearance of the optic nerve does not actuate from day to day, and unlike visual field tests, it is not dependent on patient cooperation. See: www.optic-disc.org/tutorials/glaucoma_evaluation_basics/page13.html.

In addition, compared to functional assessment, direct inspection of the optic disc seems to have the highest accuracy. The study in Fernandez-Granero *et al.* (2017) also showed in particular that cup-to-disc ratio (CDR), see Figure 2.9, is an important parameter for glaucoma detection. It is an important clinical indicator of glaucoma (Damms and Dannheim, 1993), and measures the ratio between the vertical heights of the optic cup against the vertical optic disc height. Currently, CDR assessment is performed manually by ophthalmologists to gauge and monitor optic cupping size and degeneration of the optic nerve head in glaucoma. In this thesis, the objective is to design algorithm approaches to directly assess the optic nerve changes in retinal fundus images.

2.4.5 GLAUCOMA ONH EVALUATION

Digital color fundus images are a popular imaging modality to diagnose glaucoma. A number of features can be extracted from digital fundus images to measure the damage of the optic nerve (ONH) due to glaucoma. Commonly used imaging risk factors to diagnose glaucoma include optic cup-to-disc ratio, parapapillary atrophy, disc hemorrhage, neuroretinal rim notching, neuroretinal rim thinning, inter-eye asymmetry, and retinal fiber layer defect (Monica *et al.*, 2013).

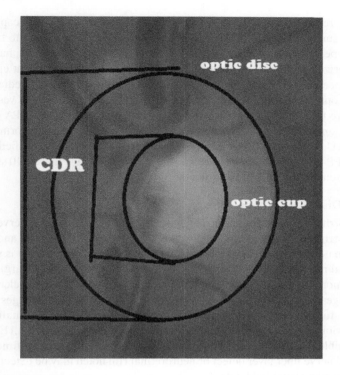

FIGURE 2.9 CDR The vertical cup-to-disc ratio is used as the clinical measure in assessing glaucoma.

2.4.5.1 Cup-to-Disc Ratio

Optic disc cupping is one of the most important signs of glaucoma (Damms and Dannheim, 1993). It happens due to cup enlargement, and is defined as the ratio of the cup diameter over the disc diameter. The optic disc (OD) has an orange-pink rim with a pale center called the optic cup, which is a neural retinal tissue that is normally white in color.

With the progression of glaucoma, more optic nerve fibers die and the OC becomes larger with respect to the OD, which corresponds to an increased CDR value. For a healthy subject, the CDR value should be around 0.2–0.3. A CDR test applied on the retinal fundus images would increase the degree of accuracy in glaucoma diagnosis (Dnyaneshwari *et al.*, 2014). CDR can be measured manually by marking the optic disc and optic cup boundaries. However, this method is quite subjective, and largely dependent on the experience and expertise of the ophthalmologists; it is both time-consuming and dependent on inter-observer variability. Thus, an automatic CDR measurement system is highly desirable.

2.4.5.2 Parapapillary Atrophy

Parapapillary atrophy (PPA) is another important risk factor that is associated with glaucoma (Jonas *et al.*, 1992). PPA is the degeneration of the retinal pigment epithelial layer, photoreceptors, and, in some situations, the underlying choroid capillaries in the region surrounding the optic nerve head. PPA can be classified into two types: an alpha type and a beta type. Alpha PPA occurs within the outer or alpha zone and is characterized by hyper- or hypo-pigmentation of the retinal pigment epithelium. Beta PPA occurs within the inner or beta zone, which is the area immediately adjacent to the optic disc, and is characterized by visible sclera and choroid vessels. PPA occurs more frequently in glaucomatous eyes, and the extent of beta PPA correlates with the extent of glaucomatous damage, particularly in patients with normal tension glaucoma (Uchida *et al.*, 1998). The development of PPA can be classified into four stages: no PPA, mild PPA, moderate PPA, and extensive PPA. Figure 2.10 shows how these different stages of PPA look like on fundus images.

2.4.5.3 Disc Hemorrhage

Disc hemorrhage is a clinical sign that is often associated with optic nerve damage. Disc hemorrhage is detected in about 4%–7% of eyes with glaucoma and is rarely observed in normal eyes. The hemorrhage is usually dot-shaped when it is within the neuroretinal rim and flame-shaped when it is on or close to the disc margin. Flame-shaped hemorrhages within the retinal nerve fiber layer that cross the sclera ring are highly suggestive of progressive optic nerve damage. Disc hemorrhages are most commonly found in the early stages of normal tension glaucoma, usually located in the inferior or superior temporal disc regions, as shown in Figure 2.11. They are usually visible for 1–12 weeks after the initial bleeding. At the same time, a localized retinal nerve fiber layer defect or neuroretinal rim notch may be detected, which corresponds to a visual field defect.

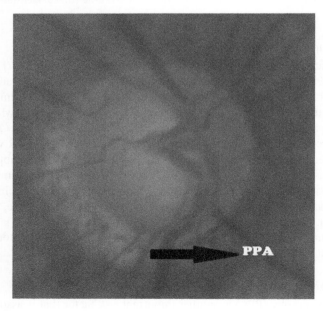

FIGURE 2.10   Grading of PPA according to scale, occurs more frequently in glaucomatous eyes.

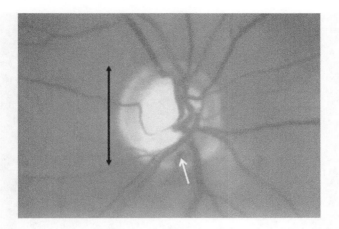

FIGURE 2.11   Disc hemorrhage in the inferior temporal side indicating glaucoma (Ravi *et al.*, 2011).

2.4.5.4   Notching

Neuroretinal rim notching, also known as the focal enlargement of the optic cup, is the focal thinning of the rim, which is a form of structural damage to the glaucomatous optic disc (Allingham, 2005). Disc hemorrhage and RNFL damage often develop at the edge of the focal notching. Thus, it is the hallmark of glaucomatous optic disc damages, and its presence is considered to be practically pathognomonic.

2.4.5.5   Neuroretinal Rim Thinning

The neuroretinal rim thinning associated with loss occurs in four sectors. First at the inferior temporal disc sector, and the nasal superior sector is the last to be affected (Harizman *et al.*, 2006; Jonas *et al.*, 1993). The measurement of the neuroretinal rim loss is essential for glaucoma detection and it is also a complement to the PPA detection, as the site of the largest area of atrophy tends to correspond with the part of the disc with the most rim loss (Kotecha, 2002). Figure 2.12 shows the rim widths in different sectors of the optic disc.

2.4.5.6   Inter-Eye Asymmetry

Inter-eye asymmetry of optic disc cupping is useful in glaucoma detection, because one eye is usually worse than the other in glaucomatous patients. In contrast, only about 3% of normal individuals have such asymmetry. Thus, inter-eye optic disc cupping asymmetry is a good sign of glaucoma.

2.4.5.7   Retinal Nerve Fiber Layer Defect

The RNFL appears as bright fiber bundle striations which are unevenly distributed in normal eyes. The fiber bundles can be observed most easily in the inferior temporal sector, followed by the superior temporal sector, the superior nasal sector, and finally the inferior nasal sector. They are rarely visible in the temporal and nasal regions. RNFL defects are associated with visual field defects in the corresponding

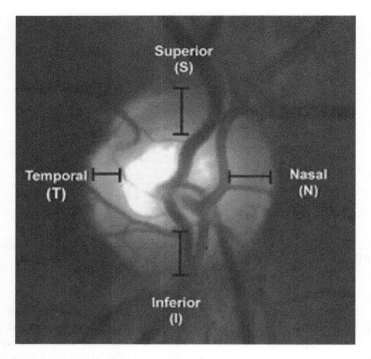

FIGURE 2.12   Shows rim widths in the inferior, superior, nasal, and temporal sectors, also indicators of the glaucoma disease based on the ISNT rule (Linda, 2017).

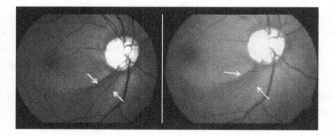

FIGURE 2.13   Images show examples of RNFL wedge-shaped defect (Thomas *et al.*, 2011).

hemifield. When RNFL defect exists, there would be dark areas in the bright striations on the fundus image. The RNFL defects are usually wedge-shaped and are commonly seen in both hypertension and normal pressure glaucoma. Figure 2.13 shows examples of the RNFL defect.

2.5   MOTIVATION OF FUNDUS IMAGE PROCESSING

The fundus represents the bottom or base of anything. The ocular fundus is the inner part of the eye made up of the Sensory Retina, the Retinal Pigment Epithelium, Bruch's Membrane, and the Choroid.

The ocular fundus can be noticed in the reflection of the flash attachment on the camera as red eyes. In ocular diseases, if fundus photography is needed for diagnostic purposes, the pupil is dilated with eye drops and a special camera called a fundus camera is used to focus on the fundus.

The fundus images can be tremendous, indicating the optic nerve through which visual "signals" are transmitted to the cerebrum and the retinal vessels which supply sustenance and oxygen to the tissue set against the red-orange shade of the color epithelium (Ophthalmic Photographers Society, 2019)

A color retinal fundus camera is widely used to photograph the retina, which contains the main glaucoma risk factors like CDR, rim, PPA, and ISNT rule, all these factors appear at the optic nerve head and its surrounding area. In addition, the fundus camera is a low-cost, stable technology which is widely used and can show the true color of the retina.

2.5.1 Digital Fundus Image Processing

The mathematical background for digital fundus image processing will be explained in three main steps:

Step (1) – Pre-processing: Image Enhancement.
Step (2) – Processing: Image Segmentation.
Step (3) – Post-processing: Features Extraction and Selection.

Software packages such as MATLAB® provide powerful means for implementing these image processing functions. The following part overviews the main image processing functions that can be used for the enhancement and segmentation of images, which will be used later in Chapter Four within the context of the methodology.

Step (1) – Pre-processing: Image Enhancement

Image enhancement is the basic process of changing a raw image so that the result would be more suitable than the original for a specific application or further processing (Gonzalez, 2008). Image enhancement is a basic step that is widely used in many applications of image processing where the subjective quality of images is important for human interpretation. These digital images can be adjusted to obtain a greater degree of accuracy in analysis. For example, you can remove noise and sharpen or brighten an image, making it easier to identify key features or disease symptoms.

There are many image enhancements such as accentuation, or sharpening an image's features (edges, boundaries), or contrasting to make a graphic display more useful for display and analysis. The main advantage of the enhancement process is that it does not increase the inherent information content in the data, but it increases the dynamic range of the chosen features so that they can be detected easily. Generally, image enhancement improves the quality of an image for the visual perception of human beings without increasing the information. It is also used for low-level vision applications (Kushwaha *et al.*, 2015).

Below are some of the major enhancements that can be applied to the fundus image, either solely or combined together with one another.

Color image processing: since the captured digital fundus image is a full-color image, it is obvious that color image processing techniques are strongly needed. For example, color splitting can be used to view and process each of the red, green, and blue channels independently from one another (Dougherty, 2009). Also, conversion between the RGB and the HIS and HSV color models are very useful in many situations; unlike the RGB color space, HSI, HSV, and Lab models decouple the color (chromaticity) and grayscale (intensity) information, HSI and HSV represent the intensity within the (I) and (V) layers, respectively, while chromaticity is represented within both the Hue (H) and Saturation (S) layers. As for the $L*a*b$ color space, the L* represents the luminosity (intensity), while the $a*$ and $b*$ layers represent the chromaticity.

2.5.2 Channel Separation

Color digital fundus images are made up of pixels in combinations of the primary colors. The separated channel is the grayscale image of the same size as a color image, made up of just one of these primary colors. For instance, the colored image has a red, green, and blue channel, and the channel extraction is an important feature to extract some information that can be used in image processing (Gonzalez and Woods, 2008).

Contrast is a major factor in any subjective evaluation of image quality. It is created due to the difference between the luminance of the surfaces, and can be defined as the difference in visual properties that makes an object distinguishable from the background.

Contrast Enhancement: when the contrast of the captured image is too low, it is difficult to detect and isolate objects of interest like OD and OC in the case of fundus images. Therefore, brightening or darkening a low-contrast image can be achieved by stretching (spreading) the histogram of that image through approaches such as histogram equalization and histogram specification. Moreover, in order to stretch the contrast of a full-color image, it is logical to spread the color intensities uniformly, leaving the colors themselves unchanged; therefore, as mentioned in the previous paragraph, the HSI color space is ideally suitable for color histogram processing (Gonzalez and Woods, 2008).

2.5.3 Histogram Equalization

The histogram equalization technique is used for adjusting image intensities to enhance contrast by modifying the intensity distribution of the histogram. This is done by giving a linear trend to the cumulative probability function associated with the image.

The histogram equalization technique relies on the use of the cumulative probability function (Arthur, www.sci.utah.edu/~acoste/uou/Image/project1/Arthur_COSTE_Project_1_report.html). Let f be a given image represented as amr by mc matrix of integer pixel intensities ranging from 0 to $L - 1$. L is the number of intensity level equal to 256. Let p denote the normalized histogram with a bin for each possible intensity.

$$P_n = \frac{\text{number of pixels with intensity } n}{\text{total number of pixels}} \quad n = 0,1,\ldots L-1. \tag{2.1}$$

The image after histogram equalization g will be defined by:

$$g_{i,j} = \text{floor}\left((L-1)\sum_{n=0}^{f_{i,j}} p_n\right) \tag{2.2}$$

Where floor () rounds down to the nearest integer. This is equivalent to transforming the pixel intensities, k off by the function:

$$T(k) = \text{floor}\left((L-1)\sum_{n=0}^{k} p_n\right) \tag{2.3}$$

The motivation for this transformation comes from thinking of the intensities of f and g as continuous random variables X, Y on [0, L – 1] with Y defined by:

$$Y = T(X) = (L-1)\int_{o}^{x} px(x)dx \tag{2.4}$$

Where px is defined as the probability density of f and T is the cumulative distributive of X multiplied by (L – 1). Assume for simplicity that T is differentiable and invertible. It can then be shown that Y defined by $T(X)$ is uniformly distributed on [0, L – 1], namely that (Mandar and Meghana, 2015):

$$py(\mathcal{Y}) = \frac{1}{L-1} \tag{2.5}$$

$$\int_{0}^{y} py(z)dz = \text{probability that } 0 \leq Y \leq y$$

$$= \text{probability that } 0 \leq X \leq T^{-1}(y) \tag{2.6}$$

$$= \int_{0}^{T-1} px(y)dy$$

$$\frac{d}{dy}\left(\int_{0}^{y} py(z)dz\right) - pY(y) = px\left(T^{-1}(y)\right)\frac{d}{dy}\left(T^{-1}(y)\right) \tag{2.7}$$

2.5.4 Filtering

Filtering is a technique for modifying or enhancing digital images. Hence filtering is a neighborhood operation determined by applying some algorithms to the values of the pixels in the neighborhood of the corresponding input pixel.

Spatial linear filters (e.g., mean filter), as well as non-linear filters (e.g., median filter) are used for image smoothing and noise removal. Conversely, first derivative filters (e.g., Prewitt and Sobel filters), as well as second derivative filters (e.g., Laplacian filter) are used for image sharpening (i.e., deblurring), as well as the detection of edges, lines, and points. Moreover, an image can be also filtered in the frequency domain in order to smoothen and sharpen it by using low-pass filters and high-pass filters, respectively (Gonzalez and Woods, 2008).

Common noise types including salt-and-pepper noise, Gaussian noise, and speckle noise can all be cleaned by using spatial filtering techniques.

a. Median filtering: is a common, non-linear method for noise filtering that uses a special methodology, not by using convolution to process the image with a kernel of coefficients, but, in each position of the kernel frame, a pixel of the image contained in the frame will become the output pixel located at the coordinates of the kernel center. The kernel frame focused on the centered pixel (m, n) of the original image, to calculate the median value of the pixels. Then, the pixel at the coordinates (m, n) of the output image is set to this median value. Generally, median filters differ from the mean filter in the smoothing characteristics; another characteristic for the median filter is that features smaller than half the size of the median filter kernel are completely removed by the filter, and large discontinuities, such as edges and large changes in image intensity, are not affected in terms of gray-level intensity, but their positions may be shifted by a few pixels. These non-linear features of the median filter remove specific types of noise, such as "salt-and-pepper noise" that may be filtered completely from an image without attenuation of significant edges or image characteristics (Raman, 2009).

The Median filter is almost always used to remove salt-and-pepper noise using the median value. If there are an even number of values, the median is the mean of the middle two. A median filter is an example of a non-linear spatial filter, by the equation:

$$f(x, y) = \text{median}\{g(s,t)\}, s, t \in Sxy \tag{2.8}$$

Let Sxy represent a set of rectangular sub-images with window size $(m \times n)$ centered at (x, y).

Median Filtering Algorithm:

$$\text{Allocate output Pixel Value} \left[\text{image width} \right] \left[\text{image height} \right]$$

$$\text{Allocate window} \left[\text{window width} * \text{window height} \right]$$

$$\text{edgex} := \left(\text{window width} / 2 \right) \text{rounded down}$$

$$\text{edgey} := \big(\text{window height} / 2\big)\text{rounded down}$$

$$\text{for } x \text{ from edgex to image width} - \text{edgex}$$

$$\text{for } y \text{ from edgey to image height} - \text{edgey}$$

$$i = 0$$

$$\text{for } fx \text{ from 0 to window width}$$

$$\text{for } fy \text{ from 0 to window height}$$

$$\text{window } [i] := \text{input Pixel Value}\big[x + fx - \text{edgex}\big]\big[y + fy - \text{edgey}\big]$$

$$i := i + 1$$

$$\text{sort entries in window}$$

$$\text{output Pixel Value } [x][y] := \text{window}\big[\text{window width} * \text{window height} / 2\big]$$

b. Adaptive filtering: this is a digital filter that has self-adjusting characteristics. Its methodology depends on adjusting the filter coefficients automatically to adapt the input image via an adaptive algorithm. Adaptive filters are a class of filters which change their characteristics according to the values of the grayscales under the mask. This is implemented by applying a function to the gray values under the mask.

c. Mean filter: this is a simple sliding-window spatial filter that replaces the center value in the window with the average (mean) of all the pixel values in the window. The window, or kernel, is usually square, but it can be any shape, as mentioned in www.markschulze.net/java/meanmed.html, the filter can represent by the equation:

$$F(x,y) = \frac{1}{mn} \pounds g(s,t) \tag{2.9}$$

d. Guided filter: the guided filter performs edge-preserving smoothing on an image, using the content of a second image, called a guidance image, to influence the filtering. The guidance image used in the filtration can be the same image, a different version of the image, or another image. A guided image filter takes into account the statistics of a region in the corresponding spatial neighborhood of the guidance image when calculating the value of the output pixel. If the guidance is the same as the image to be filtered, the structures of the output image are the same as the original image and the guidance image. When the guidance image is different, the filtered image will follow it and affect the original image structure. This effect is called structure transference.

The filtering output image is a linear transformation of the guidance image. This filter's main characteristics include edge-preserving smoothing (like the bilateral filter), but it does not suffer from the gradient reversal artifacts related to the matting Laplacian matrix, and it can be used in other applications beyond the scope of "smoothing." Moreover, the guided filter has an exact algorithm for both grayscale and color images. The guided filter shows an accurate performance in terms of both quality and efficiency in a great variety of applications, such as noise reduction, smoothing/enhancement, HDR compression, image feathering, haze removal, and joint up sampling (K. Daniilidis *et al.*, 2010).

e. Gaussian filter: the guided filter is a well-known local filter for its edge-preserving properties and low computational complexity. It is ideal for those starting to experiment with filtering. Thus, its design can be controlled by changing just one variable which is the variance. The function is defined as:

$$G(x,y) = \frac{1}{2\pi\sigma^{2^e}} - \frac{x^2 + 2_y}{2\sigma^2} \tag{2.10}$$

The value of the sigma (the variance) corresponds inversely to the amount of filtering, smaller values of sigma mean that more frequencies are suppressed and vice versa (Kou *et al.*, 2015).

Generally, it is a good general-purpose filter, and it is used for the separation of the roughness and waviness components from surfaces. Both roughness and waviness surfaces can be separated in one single filtering procedure with minimal phase distortion. The used weighting function of a real filter is the Gaussian function (Richard, 2014).

2.5.5 Morphological Processing

Morphological operations methodology is based on combining an image with a structuring element in a 3×3 matrix. They process an image pixel by pixel according to the neighborhood pixel values. Morphological operations are often applied to binary images, although techniques are available for gray-level images (Jonathan, 2005).

- The morphological operations can be applied for image enhancement techniques and image segmentation techniques. The main operations of morphological processing are dilation and erosion, in which dilation leads to a thickening of the original object.
- Dilation mainly increases the sizes of objects to fill the holes and broken areas, and connecting areas that are separated by spaces smaller than the size of the structuring element. For different types of images, such as grayscale images, dilation increases the brightness of objects by taking the neighborhood maximum when passing the structuring element over the image. But in binary images, dilation connects areas that are separated by spaces smaller than the structuring element and adds pixels to the perimeter of each image object.
- The erosion is an inverse procedure to the dilation in which an object is thinned, as it decreases the sizes of objects and removes small anomalies by subtracting objects with a radius smaller than the structuring element.

For grayscale images, erosion reduces the brightness and size of bright objects on a dark background by taking the neighborhood minimum when passing the structuring element over the image, and for binary images, erosion completely removes objects smaller than the structuring element and removes perimeter pixels from larger image objects.

Generally, erosion and dilation can be applied together, one after another. This can be done in a reversed order as well, in order to achieve opening and closing operations, respectively. Opening can be used to remove small objects from an image while preserving the shape and size of large objects in the image; whereas closing can be used for merging narrow breaks or gaps and eliminating holes in an image (Dougherty, 2009; Gonzalez and Woods, 2008). In segmentation the morphological operation can be used for boundary smoothing and blood vessel removing.

The boundary smoothing by morphological operations are:

a. **Opening**: the opening of an image f by a structuring element s (denoted by $f \circ s$) is an erosion (erosion denoted by $\ominus$), followed by a dilation (dilation denoted by $\oplus$), then an opening $f \circ s = (f \ominus s) \oplus s$. The opening operation is used to open up a gap between objects connected by a thin bridge of pixels, and the regions that survive will be restored to their original size by dilation. The benefit of opening with a disc structuring element is that it smoothens corners from the inside, and, by doing the entire operation, it removes further noise around corners from the inside and abridges the image (Ravi *et al.*, 2013).

b. **Closing**: the closing of an image f by a structuring element s (denoted by $f \bullet s$) is a dilation, followed by erosion $f \bullet s = (f \oplus s) \ominus s$, in morphological operations, closing is used to fill holes and keep the initial region sizes unchanged. This closing operation is the opposite of the opening operation used to smoothen corners. The main objective of this operation is to smoothen the contours and maintain the shape and size of the object. A combination of both of these operations is able to get better results for detecting edges in an in-depth image (Ravi *et al.*, 2013). To improve the segmentation process, vessel removal is applied to the digital fundus image as a pre-processing step.

Vessel removal works by an opening morphological operation where it is of an image f by a structuring element s (denoted by $f \circ s$) is an erosion followed by a dilation: $f \circ s = (f \ominus s) \oplus s$. This can open up a gap between objects connected by a thin bridge of pixels. Any regions that have survived the erosion are restored to their original size by the dilation, and that makes the blood vessel dilate with its surrounded pixels and disappear. Opening morphological techniques probe an image with a small shape or template called a structuring element. This is positioned at all possible locations in the image and is compared with the corresponding neighborhood of pixels. Since the optic disc is a circle shape, the structuring element chosen was a disk with a corresponding neighborhood of eight pixels. To achieve good and fast results in vessel removal, see: www.cs.auckland.ac.nz/courses/compsci773s1c/lectures/ImageProcessing-html/topic4.htm.

Step (2) – Processing: Image Segmentation

Image segmentation is a technique used for dividing an image into multiple parts to identify specific objects or other relevant information in digital images. There are many different ways to perform image segmentation, including thresholding methods (global, Otsu's method); color-based segmentation, such as K-means clustering; transform methods, such as watershed segmentation; texture methods such as texture filters. Here, the color retinal image has been segmented: www.mathworks. com/discovery/image-segmentation.html.

Image segmentation, one of the most important aspects of image processing, is still a research area in computer vision. Image segmentation is used to extract the Region of Interest for image analysis, and the division of an image into meaningful structures. The image segmentation is a basic step in image analysis, object representation, visualization, and many other image processing tasks; thus, segmenting an image into several parts makes further processing simpler, and reduces the amount of information. Mainly, segmentation depends on several different features, such as the color or texture contained in an image. Before de-noising an image, it is segmented to recover the original image. There are several image segmentation techniques that partition the image into several parts based on image features like pixel intensity value, color, texture, etc. Many segmentation methods have been proposed and used (Manjula, 2015).

The main objective of processing the fundus image typically aims for the segmentation of the OD and OC within an image. Segmentation is an essential step prior to feature extraction and classification in a fundus image. The example methods of image segmentation are:

a. **Region-Based Methods**: region-based methods for segmentation start from the middle of an object and then "grow" outward until they meet the object's boundaries. In this method, pixels that are related to an object are grouped for segmentation, then thresholding techniques and region-based segmentation are applied for the detected area to be segmented. Region-based segmentation called Similarity Based Segmentation as it look for similar pixels. The boundaries are identified for segmentation and all pixels related to the region are taken into consideration based on the change in the color and texture (R. Yogamangalam *et al.*, 2013).

   The objective of such methods is to produce connected regions, based on similarity, that are as large as possible (i.e., produce as few regions as possible) allowing some flexibility within each region (Dougherty, 2009). For example, region-growing, as its name implies, groups the seed pixels or sub-regions into larger similar regions based on predefined criteria of growth, such as specific ranges of color (Gonzalez and Woods, 2008).

b. **Boundary-Based Methods:** the objective of such methods is to determine a closed boundary, based on differences and discontinuities, such that an inside object (e.g., optic disc) and an outside boundary (edge) can be defined (Dougherty, 2009). Some of the methods used include: Gradient operators (e.g., Prewitt and Sobel kernels), the Laplacian or Laplacian of Gaussian (LOG), Difference of Gaussians (DOG), and the canny edge detector are all methods of detecting image boundaries (edges) (Gonzalez and Woods, 2008).

Edge detection is a very essential process in the field of computer vision. Thus, image edges significantly reduce the data to be processed, but they retain essential information about the shapes of objects. A major property of boundary-based methods is the ability to extract the edge line with good orientation, but generally it is difficult to evaluate the performance of edge detection techniques. Therefore it is judged personally and separately by expertise (R. Muthukrishnan *et al.*, 2011). Edge detection is an important tool for image segmentation by transforming original images into edge images using the changes of gray levels in the image and the detection of the physical and geometrical properties of the objects, where there are three types of discontinuities in the gray level: point, line, and edges. The edge detection is a fundamental process that detects the object's boundaries and background, thus, edges are local changes in the image intensity, and, generally, they occur on the boundary between two regions to serve various applications, such as medical image processing, biometrics, etc. That makes edge detection an active area of research (A. Chaudhary *et al.*, 2013).

c. **Color-Based Methods**: in order to segment an image based on color, the segmentation technique can be carried out on either the HSI color space or the RGB color space. It is intuitive to use the HSI space because color is conveniently represented in the hue plane (H), and saturation (S) is typically used as a masking image in order to isolate further regions of interest in the hue image, while the intensity (I) is used less frequently for segmentation of color images because it carries no color information. However, better results can be obtained using the RGB color vectors by measuring and comparing the Euclidean distance between an RGB pixel and a specified color range (Gonzalez and Woods, 2008).

d. **Active Contour (Snakes)**: the main aim of active contour models or snakes is to evolve a curve in order to detect the objects in an image. For instance, starting with an initial curve around the object to be segmented within an image, the curve (i.e., the snake) is pulled toward nearby features (i.e., local minima), such as lines and edges, locating them accurately (Kass *et al.*, 1988; Chan, 2001).

e. **Matched Segmentation**: a matched filter describes the expected appearance of a desired object or region of interest for purposes of comparative modeling. Thus, the segmentation process is carried out by isolating the region (pixels) that have the least accumulated difference compared to the matched filter (Youssif *et al.*, 2008).

f. **Thresholding**: this is the simplest and most common method of image segmentation, which can be applied directly to an image or combined with other techniques like pre- and post-processing. In this technique, pixels are classified to categories based on the range of pixel values (P. Singh, 2013). For example, if you wanted to segment an image at a threshold of 128, pixels with values less than 128 would be placed in one category, and those with more would be placed in the other category. The boundaries between the different categories of this image could show up as white on the original image. This would result in a new image segmented into two parts. The thresholding algorithm has three types: global thresholding, local thresholding, and adaptive thresholding.

i) Global thresholding: this method is used when the intensity distribution of objects and background pixels are sufficiently distinct. Where a single threshold value applies to the whole image. This threshold value (called T) depends only on $f(x,y)$, and the value of T solely relates to the character of the pixels (H. Devi, 2006). There are various global thresholding techniques such as Otsu, optimal thresholding, histogram analysis, iterative thresholding, maximum correlation thresholding, clustering, Multispectral, and Multi-thresholding.

ii) Local thresholding: this method is used when there is an uneven illumination due to shadows or the direction of illumination. In such situations, local thresholding would be applicable. This technique's idea is to partition the image into $m \times m$ sub-images and then choose a threshold, if threshold T depends on both $f(x,y)$ and $p(x,y)$, it is called local thresholding. This technique divides an original image into several sub-images in order to choose different thresholds Ts for each sub-image. Then, discontinuous gray levels among sub-images must be eliminated through use of the gray-level filtering technique (P. K. Sahoo, 1998). Examples of the local thresholding techniques are statistical thresholding, 2-D entropy-based thresholding, and histogram-transformation thresholding.

iii) Adaptive thresholding: this method typically takes a grayscale or color image as its input, and, in its simplest implementation, outputs a binary image representing the segmentation; the output will be a segmented binary image. For each pixel in the input image, a threshold has to be calculated. If the pixel value is below the threshold it is then set to the background value, otherwise it is segmented as foreground. In adaptive thresholding, different threshold values for different local areas are used to separate desirable foreground objects from the background using pixel intensities (E. R. Davies, 1997). The disadvantage of this method is that it is computationally expensive and not appropriate for real-time applications. An image is easily and speedily segmented by selecting a threshold, which partitions images directly into regions based on intensity values and/or the properties of those values. Global thresholding and local thresholding are two types of thresholding that may be applied to an image; global thresholding is a segmentation technique that uses a constant threshold over the entire image, whereas local (variable) thresholding uses a variable threshold that changes over an image, based on the properties of a neighborhood (e.g., average intensities of a neighborhood) (Gonzalez and Woods, 2008).

Thresholding segmentation assumes that the intensity values are different in different regions, and, similar to the ways in which it represents the corresponding object, in fixed thresholding, the threshold value is held as a constant throughout the image, treating each pixel independently of its neighborhood.

$$\text{Fixed thresholding is of the form: } g(x, y) = \sum \begin{matrix} 0 & f(x,y){<}t \\ 1 & f(x,y) \geq t \end{matrix} \quad (2.11)$$

Where t is thresholding level.

The global thresholding algorithm used for disc and cup segmentation is:
1) Select an initial estimate for T.
2) Segment the image using T. This will produce two groups of pixels. G1 consisting of all pixels with gray-level values >T and G2 consisting of pixels with values ≤T.

Where the threshold level (T) can be chosen manually or by using automated techniques, manual threshold level selection is normally done by trial and error, using a histogram as a guide (Prasantha *et al.*, 2010).

After disc segmentation step the boundary smooth step using erosion and dilation operation, where it is operation expands or thickens foreground objects in an image. Then the border cleared to suppresses structures that are lighter than their surroundings. For the segmented disc, the algorithm tends to reduce the overall intensity level in addition to suppressing border structures (Soille, 1999). Binarization is then used to convert an intensity image into a binary image with a global threshold level. This function uses Otsu's method, which chooses the threshold to minimize the interclass variance of the black and white pixels (Otsu, 1979).

g. **Morphological Watershed**: the watershed segmentation method was proposed by Digabel and Lantuéjoul, and later improved by Beucher and Lantuéjoul (Roerdink *et al.*, 2000). This technique is a region-based segmentation approach based on geography: in other words, with regards to the landscape or geography relief that is flooded by water, watersheds are the dividing lines of the domains of attraction of rain falling over the region, it is helpful to imagine the landscape being immersed by a lake, and once the water level has reached the very best peak within the landscape, the method is stopped. As a result, the landscape is divided into regions or basins separated by dams, known as watershed lines or just watersheds.

The watershed transform is a powerful segmentation tool that aims to isolate and separate the touching objects within an image. Segmenting an image via watershed is a two-step process; first, finding the markers and the segmentation criterion used to split the regions which are most often the contrast or gradient. Second, performing a marker-controlled watershed with these two elements in order to control over-segmentation (Gonzalez and Woods, 2008) (Math Works).

In this research, the digital fundus image used is a projection of retinal structures on a two-dimensional color plane where the OD appears as a bright, circular, or elliptic region, partially occluded by blood vessels. OD segmentation is a

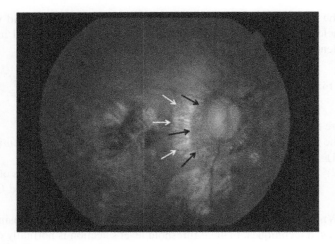

FIGURE 2.14   Show two examples of optic discs affected by atrophy (Jonas *et al.*, 2017).

challenging task mainly due to blood vessel occlusions, ill-defined boundaries, and image variations near the disc boundary due to pathological changes. Specifically, the occurrence of similar characteristics and regions (atrophy) near the disc boundary; a sample image is shown in Figure 2.14 to illustrate the above conditions.

Step (3) – Post-Processing: Features Extraction and Selection

Feature extraction is one of the most significant approaches in computer vision, hence, various applications have been improved based on image feature extraction techniques, which have attracted critical research endeavors, Such as using 2-D images' shape features to represent 3-D objects. Among the most significant and well-known features are moments, Fourier transforms, and the bag-of-words descriptor (Yue *et al.*, 2015).

After the eye fundus image has been pre-processed and the segmented features have been extracted from the internal skeleton of the recognized region, the results will be classified to distinguish or annotate the glaucoma disease with the appropriate description, and then automatically compared to the ground truth in order to evaluate the accuracy of the experimental work. This will now be explained in the following section.

2.6 FEATURE EXTRACTION BACKGROUND

The primary task of feature extraction is to take an input pattern and correctly assign it as one of the possible output classes to recognize the image as glaucomatous or healthy. This process can be divided into three general stages:

Feature extraction, selection, and classification. Feature extraction is the collection of information from a specific part of the image. It is important to the whole process since the classifier will not be able to recognize right from wrong features. Definition to choose the powerful features given by Lippman are:

Features should contain the information required to distinguish between classes, be insensitive to irrelevant variability in the input, and also be limited in number, to permit, efficient computation of discriminant functions and to limit the amount of training data required.

In glaucoma detection, feature extraction and selection are very important steps in the construction of the CAD system. In this process, relevant features are extracted from the cup, disc, and RNFL to form glaucoma features. These features are then used by classifiers to recognize the input image compared with the background image. Thus, feature extraction is the process to extract a set of features from the raw data to maximize the recognition rate with the least amount of elements and to generate a similar feature set for the recognition of new images (Gaurav *et al.*, 2014).

In digital fundus images, feature extraction is classified based on the types of features. The type of features are divided into two groups: morphological and non-morphological (Anindita Septiarini *et al.*, 2015). The morphological features are geometric parameters extracted from the image after segmentation process like RNFL, PPA, and CDR, disc and cup area, CDR, and neuro-retinal rim area. The non-morphological features are whole image features, which means they are extracted from the existing image (image-based features). Color, shape, and texture features are the type of the features that are extracted from the whole image to represent the characteristics of the glaucoma disease. The color feature can be used to recognize the cup, neuro-retinal rim, and PPA features. The blood vessels, neuro-retinal rim, and PPA can be extracted using shape features, while RNFL can only be extracted by the texture feature.

Anum *et al.* (2016) proposed a method that introduces a suspect class in automated diagnosis based on structural and non-structural features, and evaluates this algorithm using a local database containing 100 fundus images. This system is designed to classify glaucomatous cases from non-glaucomatous cases, and the motivation behind introducing the suspect class is to ensure a high sensitivity of the proposed system. The system's results achieved an average sensitivity and specificity of 100% and 87%, respectively.

An example of a non-morphological feature was presented by Rama *et al.* (2012), which discussed a system for the automated detection of normal and glaucomatous cases using higher-order spectra (HOS), trace transform (TT), and discrete wavelet transform (DWT) features from the whole fundus image. Then the extracted features were classified using the support vector machine (SVM). In this work, the SVM classifier achieved an accuracy of 91.67%, and a sensitivity and specificity of 90% and 93.33%, respectively. Furthermore, they proposed an algorithm, called the Glaucoma Risk Index (GRI), which was composed of HOS, TT, and DWT features to aid clinicians in making a faster glaucoma diagnosis during the mass screening of normal/glaucomatous images

In an example of a morphological feature in the study by Xu *et al.* (2007), the disc and cup were segmented using the free-form deformable model (snake) technique, then the boundary was extracted based on the combination of information from the smoothness, gradient, and depth as a modification to the original snake technique. The method was tested in 100 retinal fundus images having both normal and glaucomatous cases from the National University Hospital. The results had an accuracy for boundary detection of 94%.

The features used in this research combined the morphological and non-morphological features by extracting the shape, color, and texture features from the segmented OD, OC, and RNFL parts.

2.7 SHAPE FEATURES

Shape-based features extraction depends on the measuring of similarity between shapes represented in images based on various features, such as simple geometric features that can be used to describe shapes. Generally, the geometric features can distinguish shapes with huge differences; that makes it so that filters can eliminate false hits or be combined with other shape features to recognize shapes.

Shape features are features used to describe an object, using its most important characteristics. They can be described by many parameters, such as the center of Gravity/centroid, the axis of least inertia, digital bending energy, eccentricity, circularity ratios, elliptic variance, rectangularity, convexity, solidity, Euler number, profiles, and whole area ratio.

Shape feature extraction has been used in various applications:

a. Shape retrieval: searching for all shapes in a large database of shapes.
b. Shape recognition and classification: to match shapes with a model (or to the one with the most similarity).
c. Shape alignment and registration: change a shape to the best match, either as a whole or in part.
d. Shape approximation and simplification: reconstruct a shape from fewer elements based on the similarity to the original (Mingqiang, 2008).

In this research, several properties describe the shape for each disc and cup within an image via a built-in function in MATLAB® (region prop): https://octave.sourceforge.io/image/function/regionprops.html.

Region props compute Area, Euler Number, Centroid, Filled Area, Filled Image, Extent, Major Axis Length, Minor Axis Length, Solidity, and Perimeter; these measurements are applied for disc and cup masks. For measuring these features, a binary image obtained the view as a binary function:

$$f(x,y) = \begin{cases} 1 & \text{if } (x,y) \in D \\ 0 & \text{otherwise} \end{cases} \tag{2.12}$$

2.7.1 THE CENTROID

The centroid determined the object's center points, which can be illustrated by this formula:

Centroid $(\vartheta x, \vartheta y)$ is:

$$\begin{cases} \vartheta x = \dfrac{1}{N} \sum_{i}^{N} 1xi \tag{2.13} \\ \\ \vartheta y = \dfrac{1}{N} \sum_{i}^{N} 1yi \tag{2.14} \end{cases}$$

Where N represents the number of point in the shape.

2.7.2   ECCENTRICITY

Eccentricity is the measure of the aspect ratio, its ratio of the length of the major axis to the minor axis, which calculated by the principal axes method or a minimum bounding rectangular box.

2.7.3   SOLIDITY

The ratio of the pixels in the convex hull is found in the region returned as a scalar. This is computed as Area/Convex Area.

Solidity describes the extent to which the shape is convex or concave (Annesha, 2013), and it is defined as:

$$\text{Solidity} = A_s/H \tag{2.15}$$

Where, As represents the shape area and H is the convex hull area of the shape; therefore, the solidity of a convex shape is equal 1.

2.7.4   AREA

The actual number of pixels in the region, can be calculated as the equivalent diameter of a circle with the same area as the region returned using the formula:

$$\text{Area} = (4 * \text{Area}/pi) \tag{2.16}$$

2.7.5   MAJOR AXES

The length of the major axis of the ellipse that has the same normalized second central moments as the specific region in pixels.

2.7.6   MINOR AXES

The length of the minor axis of the ellipse that has the same normalized second central moments as the specific region in pixels.

2.7.7   EXTENT

The extent is the ratio of pixels in the segmented region to pixels in the total bounding box, as it is computed using the area divided by the area of the bounding box.

2.7.8   PERIMETER

The perimeter is the length of the entire outside boundary of the segmented region, by calculating the distance between each adjoining pair of pixels around the boundary of the region.

2.8 COLOR FEATURES

Thus the color features are one of the most important elements in recognizing images and they can be easily recognized by the human eye. Color features are fundamental characteristics of the content of images. They provide important information and are very useful for image recognition. Are measures that characterize color distribution in an image in order to compare between images based on color features, since color moments calculate both shape and color information they are a good features to use under changing lighting conditions. The greatest advantage of using color moments comes from the fact that there is no need to store the complete color distribution (Stricker and Orengo, 2015).

Not only do these color features extract the color distribution, they also extract the spatial information of pixels in images. In addition, they have been used in research such as Annesha and Joydeep (2013). Their paper proposed a new multi-feature image clustering technique with a high accuracy level, using color moment features from an image before combining them with the features of a histogram analysis. Finally, they used a canny edge detection technique to combine all features in a matrix, and performed a clustering algorithm to cluster data and achieved an accuracy of 90.5%.

Stricker and Orengo (2009) used three central moments of an image's color distribution. These were Mean, Standard Deviation, and Skewness, which calculated to the three channels (Red, Green, and Blue), therefore, they were characterized by 9 moments: 3 moments for each 3 color channels. We will define the i–th color channel at the j–th image pixel as Pij. The three color moments can defined as:

2.8.1 MEAN

The first color feature can be explained as the average color in the image, and it can be calculated by the formula:

$$Ei = \sum_{j}^{N} 1 \frac{1}{N} p_{ij} \tag{2.17}$$

Where N is the number of pixels in the image pij, and the value of the j–th pixel of the image is the i–th color channel.

2.8.2 STANDARD DEVIATION

The second color feature is identified by calculating the square root of the variance of the color distribution.

$$\sigma i = \sqrt{\left(\frac{1}{N} \sum_{j=1}^{N} \left(p_{ij} - E_i \right)^2 \right)} \tag{2.18}$$

Where E_i represents the mean value for the *i–th* color channel of the image.

2.8.3 SKEWNESS

The third color feature is Skewness. It is used to measure how asymmetric the color distribution is, and give information about the shape of the color distribution. Skewness can be computed by the following formula:

$$S_i = \sqrt[3]{\left(\frac{1}{N} \sum_{j=1}^{N} \left(p_{ij} - E_i \right)^3 \right)} \qquad (2.19)$$

2.9 TEXTURE FEATURE

Texture features can describe as a repeated pattern of information of the structure with regular intervals. There are many methods that can be used to describe the main features of the textures, such as directionality, smoothness, coarseness, and regularity. The **gray-level co-occurrence matrices** measure is one of the most important measures that can be used to describe texture. In this research, two techniques were used to describe the RNFL texture.

In 1973, Haralick proposed the co-occurrence matrix and texture features, which are the most popular second-order statistical features. This method proposed two steps for texture feature extraction, the first step is computing the co-occurrence matrix and the second step is calculating texture feature based on the co-occurrence matrix and it is useful in wide range of image analysis applications.

GLCM feature method: the basic GLCM texture features extraction is the relation between two neighboring pixels in one offset as the second-order texture. The gray value relationships in a target are transformed into the co-occurrence matrix space by a given kernel mask, such as 3×3, 5×5, 7×7. To change the image space into the co-occurrence matrix space, the neighboring pixels in one (or some) of the 8 defined directions can be used; basically, 4 directions $0°$, $45°$, $90°$, and $135°$ are initially regarded, and its reverse direction (negative direction), too, which contains information about the positions of the pixels having similar gray-level values. A method to calculate the spatial relationships of pixels known as the gray level – also known as the gray-level spatial dependence matrix – characterizes the texture of an image by calculating how often pairs of pixels with specific values, which are found in a specified spatial relationship, occur in an image, creating a GLCM, and then extracting statistical measures. These measures are:

Autocorrelation, Contrast, Correlation, Cluster Prominence, Homogeneity, Cluster Shade, Difference variance, Dissimilarity, Energy, Entropy, Maximum probability, Sum of squares, Sum average, Sum variance, Sum entropy, Difference entropy, Information measure of correlation, Inverse difference, Inverse difference normalized, Inverse difference moment normalized. Some formula examples are GLCM texture features that emphasize the spatial relationships between pixels, rather than single pixel intensities. The gray-level co-occurrence matrix represents the relationships of combinations of the pixel intensities and provides second-order statistical information. Features generated by GLCM are also known as Haralick features, where extractions from RGB images are converted into grayscale images. Here, 22 features are extracted from this GLCM.

2.9.1   GLCM ALGORITHM

The feature is calculated after quantizing the image, and each sample is treated as a single image pixel. The value of the sample is the intensity of that pixel, which is then further quantized into a specified number of discrete gray levels.

1. Create the GLCM as a square matrix with the size $N \times N$, where N is the number of levels specified under quantization, using these steps:
   a. The sample under consideration for the calculation equals s.
   b. Let W be the set of samples surrounding the s, within a window centered by s of the size specified under the window size.
   c. Considering only the samples in the set W, define each element i, j of the GLCM as the number of times two samples of intensities i and j occur in a specified spatial relationship (where i and j are intensities of a value between 0 and Number of Levels-1). The sum of all the elements i, j of the GLCM represents the total number of times the specified spatial relationship occurs in W. Make the GLCM symmetric:
      i.   Make a transposed copy of the GLCM.
      ii. Add this copy to the GLCM itself. This produces a symmetric matrix in which the relationship i to j is indistinguishable from the relationship of j to i (for any two intensities of i and j). Then the sum of all the elements i, j of the GLCM will be twice the total number of times the specified spatial relationship occurs in W (once where the sample with intensity i is the reference sample, and once where the sample with intensity j is the reference sample), and for any given i, the sum of all the elements i, j with the given i will be the total number of times a sample of intensity i appears in the specified spatial relationship with another sample (https://support. echoview.com/WebHelp/Windows_and_ Dialog_ Boxes/Dialog_ Boxes/Variable _properties_dialog_box/Operator_pages/GLCM_ Texture_Features.htm).
2. Normalize the GLCM: to apply this step, one must divide each element by the sum of all elements. The elements of the GLCM may now be considered as the probabilities of finding the relationship i, j (or j, i) in W.
3. Calculate the selected feature: the selected features can be calculated using the values in the GLCM. After the GLCM matrix has extracted the 22 features used to describe these relations, an example of these equations are:

Energy feature

$$\text{Energy} = \sum_{i,j=0}^{N-1} \left(p_{ij}\right)^2$$

Entropy feature

$$\text{Entropy} = \sum_{i,j=0}^{N-1} -\ln\left(p_{ij}\right)\left(p_{ij}\right)$$

Contrast feature

$$\text{contrast} = \sum_{i,j=0}^{N-1} p_{ij}\left(i-j\right)^2$$

Homogeneity feature

$$\text{Homgeneity} = \sum_{i,j=0}^{N-1} \frac{p_{ij}}{1(i-j2)}$$

Correlation feature

$$\text{correlation} = \sum_{i,j=0}^{N-1} p_{ij} \frac{(i-\mu)(j-\mu)}{\sigma^2}$$

Shade feature

$$\text{Shade} = \text{sgn}(A)\lceil A \rceil \tfrac{1}{3}$$

Prominence feature

$$\text{Prominence} = \text{sgn}(B)\lceil B \rceil \tfrac{1}{4}$$

Where:

Pij = Element i, j of the normalized symmetrical GLCM

N = represents the number of gray levels in the image as specified by the number of levels under quantization in the GLCM.

μ = the GLCM mean (being an estimate of the intensity of all pixels in the relationships that contributed to the GLCM), calculated as:

$$\mu = \sum_{i,j-0}^{N-1} ip_{ij} \tag{2.20}$$

Note: the mean of all the pixels in the data window W, $\sigma 2$ is the variance of all pixels that contributed to the GLCM, calculated as:

$$\sigma^2 = \sum_{i,j-0}^{N-1} p_{ij}\left(i-\mu\right)^2 \tag{2.21}$$

Note: the variance of the intensities of all the pixels in the data window W https://support.echoview.com

$$A = \sum_{i,j=o}^{N-1} \frac{\left(i+j-2_\mu\right)^3 p_{ij}}{\sigma 3\left(\sqrt{2(1+C)}\right)^3} \tag{2.22}$$

2.10 TAMURA METHOD

The Tamura method describes features according to quantitative analysis given by Tamura *et al.* (1978). This method proposed six textural properties and gave descriptions common over all texture patterns in digital images to calculate Coarseness, Contrast, and Directionality features for fundus images.

2.10.1 COARSENESS

The most important texture feature that aims to identify the largest size at which a texture exists, and even at which a smaller micro texture exists. It calculates (and takes averages at) every point over neighborhoods where the linear size is in powers of 2. The average over a neighborhood with a size of $2k \times 2k$ at point (x, y) is calculated using:

$$A_k\left(x,y\right) = \sum_{i=x-2^{k-1}-1}^{x+2^{k-1}-1} \sum_{j=y2^{k-1}}^{y+2^{k-1}-1} f(i,j)/2^{2k} \tag{2.23}$$

This take the difference between pairs of averages at each point, based on non-overlapping neighborhoods on opposite sides of the point in both horizontal and vertical orientations. In the horizontal case, this is:

$$E_{k,h}\left(x,y\right) = \left| A_K\left(x+2^{k-1},y\right) - \left(x-2^{k-1},y\right) \right| \tag{2.24}$$

At each point, one then picks the best size which gives the highest output value, where k maximizes E in either direction, then calculates the average of S opt $(x, y) =$ $2k$ opt over the picture.

2.10.2 CONTRAST

This is a statistical distribution of pixel intensity defined as $\alpha4 = \mu4/\sigma4$ where $\mu4$ represents the fourth moment and $\sigma2$ is the variance. Contrast is measured by the following formula:

$$F_{\text{con}} = \frac{\sigma}{\alpha_4^{1/4}} \tag{2.25}$$

2.10.3 DIRECTION DEGREES

To calculate the direction of the gradient vector at each pixel, the direction of the vector mode is defined as:

$$\left|\Delta G\right| = \left(\left|\Delta_H\right| + \left|\Delta_v\right|\right)/2$$

$$\theta = \tan^{-1}\left(\Delta_v / \Delta_H\right) + \pi/2 \tag{2.26}$$

Where in Δ_H and Δ_V are the following two 4×4 operator variations resulting in horizontal and vertical directions by the image convolution.

$$
\begin{array}{cccccc}
-1 & 0 & 1 & 1 & 1 & 1 \\
-1 & 0 & 1 & 0 & 0 & 0 \\
-1 & 0 & 1 & -1 & -1 & -1
\end{array}
\tag{2.27}
$$

Using these steps:

1. The gradient vector of all the pixels is calculated.
2. Histogram is built for the expression of HD θ value.
3. The first range of values θ histograms was discrete, then the corresponding statistics for each bin of $| \Delta G |$ was greater than the number of pixels in a given threshold.
4. The histogram of an image exhibits a peak for a clear directional, for no apparent direction of the image is a relatively flat performance.
5. The final image can be calculated by the directional sharpness of peaks in the histogram obtained. This is defined as follows:

$$
F_{\text{dir}} = \sum_{p}^{np} \sum_{\varnothing \in \omega \rho} (\theta - \theta \rho) 2 H_D(\varnothing)
\tag{2.28}
$$

Where np is the histogram of all the peaks, p, Wp represents all peaks included in the bin, and the bin having the highest φp value (Swati and Yadav, 2015).

2.11 FEATURE SELECTION

Feature selection is the process where you automatically or manually select those features that are most relevant and the contribute the most to the prediction variable or output in which you are interested in, thus irrelevant features can decrease the accuracy of the models and make your model learn based on irrelevant features. It is the process to select the most relevant features (variables, predictors) to be used in the CAD system reconstruction, for four main reasons:

a) Simplification of the system for researchers and users (Gaurav et al., 2014).
b) Shorter training and prediction time.
c) To avoid and minimize the curse of dimensionality.
d) Enhanced accuracy by reducing over fitting and reduction of variance (Gareth et al., 2013).

The important thing to remember with regards to the feature selection technique is to remove the **redundant** or **irrelevant** data, without incurring much loss of information (Bermingham et al., 2015). Redundant or irrelevant features decrease the system's accuracy, since one relevant feature may be redundant in the presence of another relevant feature with which it is strongly correlated (Guyon et al., 2003). Feature selection techniques used when their too many features and few samples. In this research, two types of feature selection are used:

1. Sequential feature selection.
2. T-test feature selection.

2.11.1 Sequential Feature Selector

Sequential feature selection algorithms are used to reduce an initial d-dimensional feature space to a k-dimensional feature subspace where $k < d$. An approach such as sequential feature selection is a wrapper technique. This is useful if it is an embedded feature selection technique like LASSO.

This method removes or adds one feature at a time based on the classifier performance until a feature subset of the desired size k is reached. There are four different types of the Sequential Feature Selector:

1. Sequential Forward Selection (SFS).
2. Sequential Backward Selection (SBS).
3. Sequential Forward Floating Selection (SFFS).
4. Sequential Backward Floating Selection (SBFS).

The SFFS and SBFS selectors are extensions to the simpler SFS and SBS algorithms, where these floating algorithms have an additional exclusion or inclusion step to remove features once they are included (or excluded), so that a larger number of feature subset combinations can be sampled if the resulting feature subset is assessed as "better" by the criterion function after the removal (or addition) of a particular feature.

2.11.2 Sequential Forward Selection (SFS)

$$\textbf{Input}: Y = \{y1, y2, ..., yd\} \tag{2.29}$$

a. The **SFS** algorithm takes the whole d dimensional feature set as input.

$$\textbf{Output}: Xk = \{xj \mid j = 1, 2, ..., k; xj \in Y\} \tag{2.30}$$

Where $k = (0, 1, 2 ... d)$

b. SFS returns a subset of features; the number of selected features k, where $k < d$ has to be specified a priori.

$$\textbf{Initialization}: X0 = \varnothing, k = 0 \tag{2.31}$$

- The algorithm with an empty set ("null set") so that $k = 0$ (where k is the size of the subset).

Step 1 (Inclusion):

$$x+ = \text{argmax}\, J\left(xk + x\right), \text{ where } x\, Y - Xk \qquad (2.32)$$

$Xk + 1 = Xk + x + k = k + 1$

Go Step 1

An additional feature $x+$ added to the feature subset $Xkx+$ is the feature that maximizes our criterion function; that is, the feature that is associated with the best classifier performance if it is added to Xk. This final step is repeated until the termination criterion is satisfied.

Termination: $k = p$

Then features from the feature subset Xk are added until the feature subset of size k contains the number of desired features p a priori (Ferri *et al.*, 1994; Pudil *et al.*, 1994).

2.12 CLASSIFICATION

Image classification refers to an approach in computer vision that can classify an image according to its visual content. For example, an image algorithm may be designed to tell if an image contains a human figure or disease symptom.

The selected features of image representation that are generated from feature selection are used in glaucoma detection using knowledge-based models and classification methods (Dhawan and Dai, 2008). These feature and image classification techniques are as follows:

2.12.1 STATISTICAL CLASSIFICATION METHODS

The categories of these methods have an unsupervised and a supervised approach. The unsupervised method clusters the data based on its separation in the feature space, and includes K-means and fuzzy clustering. On the other hand, a supervised approach needs training data, test data, and class labels to classify the data; it also needs to include probabilistic methods like the nearest neighbor and Bayesian classifier (Dhawan and Dai, 2008).

- Supervised classification: this method classifies a set of specific classes by providing training statistics that identify each category.
- Supervised classifiers (algorithms):
  - Parallelepiped – based on range or variance of class.
  - Minimum Distance-to-Means – based on mean class.
  - Maximum Likelihood – based on probability of class membership.
  - Spectral Angle Mapper – class membership based on minimum difference from the n-dimensional spectral vectors of the classes.
- Unsupervised classification: the raw spectral data are grouped first, based on the statistical structure of the data. Then, the statistical cluster will be labeled into the appropriate categories. However, the analyst has the task of interpreting the many classes that are generated by the many classes that

are generated by the unsupervised classification algorithm, which is known as clustering, because it is based on the natural groupings of pixels in image data in feature space (David, 2015).

2.12.2 Rule-Based Systems

This system analyzes the feature vector using multiple sets of rules that are designed to test specific conditions in the feature vector database to set off an action. The rules consist of two parts: condition premises and actions, which are generated based on expert knowledge to decide the action when the conditions are satisfied, where the action taken will be a part of the rule that could change or a labeling of the feature vector based on the analysis.

Usually, a rule-based system consists of three sets of rules: supervisory or strategy rules, a focus of attention rules, and knowledge rules. The supervisory or strategy rules control the analysis process and provide the control actions, including starting and stopping action.

The strategy rules determine which rules would be tested during the analysis process. The focus-of-attention rules provide specific features within the analysis process by accessing and extracting the information or features from the database. Thus, the rules contain the information from the database and the activity center where the implementations of knowledge rules are scheduled. Finally, the knowledge rules analyze the information related to the required conditions, and then execute an action that changes the output database (Dhawan and Dai, 2008). Examples of **supervised classification** are:

2.12.3 Support Vector Machine (SVM)

The SVM is a supervised machine learning algorithm used for both classification and regression problems, but mostly for classification. The SVM plots each feature value as a point in n-dimensional space. Then, classification is performed by finding the hyper-plane that differentiates between the two classes.

Support Vector Machines are very powerful classification algorithms, used in conjunction with random forest and other machine learning tools to obtain a very high degree of predictive power. You will find these algorithms very useful in solving some of the medical imaging classifications you come across. In addition to performing linear classification, SVMs can efficiently perform a non-linear classification using what is called the kernel trick, implicitly mapping their inputs into high-dimensional feature spaces (Sunil Ray, 2017).

2.12.4 *k*-Nearest Neighbors Algorithm (*k*-NN)

The *k*-nearest neighbor's algorithm (*k*-NN) is a non-parametric method used for classification. The input consists of the *k* closest training examples in the feature space. The *k*-NN can be used for classification or regression:

1. In *k-NN classification*, the output is classified by a majority vote of its neighbors, as assigned to the class most common among its *k* nearest neighbors

(k is a positive integer, typically small), when $k=1$, the object is simply assigned to the class of that single nearest neighbor.

2. Where k is a user-defined constant, and an unlabeled vector (test point) is classified by assigning the label which is most frequent among the k training samples nearest to that test point.

3. The distance metric used for continuous variables is the Euclidean distance metric, and for discrete variables it is the overlap (or hamming distance) metric. For gene expression microarray data, the k-NN can be employed with correlation coefficients like Pearson and Spearman (Jaskowiak, 2011). The classification accuracy of k-NN can be improved if the distance metric is learned with specialized algorithms such as Large Margin Nearest Neighbor or Neighborhood components analysis.

Distance Functions

$$\text{Euclidean} \sqrt{\sum_{i=1}^{k}\left(x_i - y_i\right)^2} \tag{2.33}$$

$$\text{Manhattan} \left|x_i - y_i\right| \tag{2.34}$$

$$\text{Minkowski} \left(\sum_{i=1}^{k}\left(\left|x_i - y_i\right|\right)^q\right)^{1/q} \tag{2.35}$$

Note: the three distance measures are only valid for continuous variables, for categorical variables the Hamming distance must be used; it must also be used when there is a mixture of numerical and categorical variables in the data set.

Hamming Distance

$$D_H = \sum_{i-1}^{k} x_i - y_i \tag{2.36}$$

$$x = y \Rightarrow D = 0$$

$$x \neq y \Rightarrow D = 1$$

The KNN algorithm's (Adi Bronshtein, 2017) steps are:

1. A test point k is specified, along with a new sample.
2. The k entries selected in the database which are closest to the new sample.
3. The most common classification found of these entries is the classification we give to the new sample.

KNN features:

    i. KNN stores the training data set, and does not learn any model.

    ii. KNN makes predictions by comparing the similarity between an input sample and each training instance.

2.12.5   SOME ADVANTAGES AND DISADVANTAGES OF KNN

2.12.5.1   Advantages

1. For non-linear data, no assumptions about data are that good.
2. It is a simple, easy algorithm to understand.
3. It has a high accuracy, but it is not competitive in comparison with better supervised learning classifiers.
4. It is useful for classification and regression.

2.12.5.2   Disadvantages

1. High memory requirement for saving the results.
2. Stores all training data, making it computationally expensive.
3. Prediction stage might be slow (with big N) and take time.
4. Sensitive to irrelevant features and the scale of the data, which reduces its accuracy.

2.12.6   ENSEMBLE LEARNING

Zhi-Hua Zhou definition:

> Ensemble learning is a machine learning paradigm where multiple learners are trained to solve the same problem. In contrast to ordinary machine learning approaches which try to learn *one* hypothesis from training data, ensemble methods try to construct a *set* of hypotheses and combine them to use.

Ensemble algorithms are used to obtain better predictive performance than could be obtained from using any one classifier alone. Unlike a statistical ensemble which is concrete by finite set of features, but allow for some flexible ones to exist among those features.

2.12.6.1   Bootstrap Aggregating (Bagging)

Bootstrap aggregating (also known as bagging) is designed to improve the stability and accuracy of machine learning algorithms used in statistical classification/regression. It is useful for reducing variance, avoiding over fitting, and applying to decision tree methods, and it can be used with any type of method.

The Bootstrap Aggregating Technique works via generating a training set of size n, bagging m new training sets, each of size n', by sampling from D uniformly and with replacement. If $n'=n$, then for large n the set to have a fraction $(1-1/e)$ ($\approx63.2\%$) of the unique examples of D, the rest being duplicates (Aslam *et al.*, 2007). This type of sample is known as a bootstrap sample. The m models are fitted using the above m

bootstrap samples and merged with the averaging output for regression or voting for classification (Shinde *et al.*, 2014).

On the other hand, it can mildly improve the performance using methods such as K-nearest neighbors (Breiman, 1996).

2.12.6.2 Boosting

Boosting techniques work based on training each new model instance to emphasize the training instances that previous models misclassified. Sometimes, boosting has been shown to have better accuracy than bagging, but it can also tend to overfit the training data. Therefore, the most common implementation of Boosting is AdaBoost. Nowadays, some newer algorithms are reported to achieve better results.

2.12.7 Logistic Regression (Predictive Learning Model)

This is a statistical method for analyzing a data set in which there are one or more independent variables that determine an outcome. The outcome is measured with a dichotomous variable (in which there are only two possible outcomes). The goal of logistic regression is to find the best fitting model to describe the relationship between the dichotomous characteristics of interest (dependent variable=response or outcome variable) and a set of independent (predictor or explanatory) variables. This is better than other binary classifications like nearest neighbor since it also quantitatively explains the factors that lead to classification.

2.12.8 Decision Trees

A decision tree builds classification or regression models in the form of a tree structure. It breaks down a data set into smaller and smaller subsets while an associated decision tree is incrementally developed at the same time. The final result is a tree with decision nodes and leaf nodes. A decision node has two or more branches, and a leaf node represents a classification or decision. The topmost decision node in a tree which corresponds to the best predictor is called the root node. Decision trees can handle both categorical and numerical data.

2.12.9 Random Forest

Random forests or random decision forests are an ensemble learning method for classification, regression, and other tasks, which operate by constructing a multitude of decision trees at training time before outputting the class that is the mode of the classes (classification) or mean prediction (regression) of the individual trees. Random decision forests correct for decision trees' habit of over fitting to their training set.

2.12.10 Neural Network

A neural network consists of units (neurons) that are arranged in layers, which convert an input vector into some output. Each unit takes an input, applies a (often non-linear) function to it, and then passes the output to the next layer. Generally, the networks are defined to be feed-forward: a unit feeds its output to all the units on the next layer, but there is no feedback to the previous layer. Weightings are applied to the signals passing from one unit to another, and it is these weightings which are tuned in the training phase to adapt a neural network to the particular problem at hand (Mandy, 2017).

2.13 CLASSIFICATION IMBALANCED

Imbalanced features are the problem where one class outnumbers another by a large proportion; it is found more frequently in binary classification problems than in multi-level classification problems. Thus, the problem with imbalanced class's resultant is ignore cases of the minority class as noise and predict the majority class. To solve this problem, there are numerous correction methods which correct the imbalanced classification problem. These methods can generally be divided into cost- and sampling-based approaches.

Thus, a data set that exhibits an unequal distribution between its classes is considered to be imbalanced, and this problem affects the classifier's performance; these are some methods used to solve the problem and result in an improved classification performance.

1. Under-sampling.
2. Oversampling.
3. Synthetic Data Generation.
4. Cost-Sensitive Learning.

2.13.1 UNDER-SAMPLING

This method is used with a majority class to reduce the number of observations from it and make the data set balanced. It is best used on a huge data set to reduce the number of training samples, and it also helps to improve run time and storage troubles. Under-sampling techniques are:

2.13.1.1 Random Under-Sampling For the Majority Class

Random under-sampling is a simple under-sampling technique, which under-samples the majority class randomly and uniformly, leading to the loss of information. But in the case of the majority class being near to the minority class, this method might yield good results.

2.13.1.2 Near Miss

To solve the information losing problem, "near neighbor" methods have been proposed with algorithms that calculate the distances between all instances of the majority class and the minority class. Then k instances of the majority class that have the smallest distances to those in the minority class are selected, and the n instances in the minority class, the "nearest" method will result in $k*n$ instances of the majority class.

The first one is the "NearMiss-1" method, which selects samples of the majority class that have their average distances closest to three instances of the minority class that are the smallest. The second method is "NearMiss-2," which uses the three farthest samples of the minority class. The last one is "NearMiss-3," which selects a given number of the closest samples of the majority class for each sample of the minority class.

2.13.1.3   Condensed Nearest Neighbor Rule (CNN)

The Condensed Nearest Neighbor method removes redundant instances that should not affect the classification accuracy of the training set. The algorithm starts with two blank data sets named A and B. Initially, the first sample is placed in data set A, and the rest of the samples are placed in data set B. After that, one instance from data set B is scanned by using data set A as the training set. Finally, if a point in B is misclassified, it is transferred from B to A. This process will continue until no instances are transferred from B to A.

2.13.1.4   Tomek Links

The Tomek Links are an effective method that considers the samples around the borderline. This method depends on observing the instance and calculating the distance between them; for example, instances a and b belonging to different classes are separated by the distance $d\,(a,\,b)$, the pair $(a,\,b)$ is called a Tomek link if there is no instance c, such that $d\,(a,\,c) < d\,(a,\,b)$ or $d\,(b,\,c) < d\,(a,\,b)$. Then the instances participating in the Tomek links are either borderline or noisy so they are both removed (Dataman, 2018).

2.13.1.5   Edited Nearest Neighbor Rule (ENN)

The Edited Nearest Neighbor Rule (ENN) methodology is based on removing any instance whose class label is different from the class of at least two of its three nearest neighbors. The concept is to remove the instances from the majority class around the borderline of different classes based on the nearest neighbor (NN) to increase the classification accuracy of minority instances and minimize the bias of the majority instances (Dataman, 2018).

2.13.1.6   Neighborhood Cleaning Rule

The Neighborhood Cleaning Rule (NCL) modifies the majority and minority instances separately when sampling the database. The NCL uses the ENN classifier to remove the majority class. For each instance in the training set, it searches the three nearest neighbors and goes through the following algorithm:

1. Search for the majority: if the instance belongs to the majority class and the classification given by its three nearest neighbors is the opposite of the class of the chosen instance, it will be removed.
2. Search for the minority: if the chosen instance belongs to the minority class and is misclassified by its three nearest neighbors, then the nearest neighbors that belong to the majority class will be removed.

2.13.1.7   Cluster Centroids

This technique under-samples the majority class based on replacing the cluster of majority instances. This method uses the K-mean algorithms to find the clusters, then uses the cluster centroids of the N clusters as the new majority samples (Dataman, 2018).

2.13.2   OVERSAMPLING

This method works with the minority class to replicate the observations from the minority class and balance the data. This is known as up-sampling. Some oversampling techniques are:

2.13.2.1   Random Oversampling for the Minority Class

The random oversampling technique randomly replicates the minority class instance, and it is known to increase the likelihood of over fitting. On the other hand, the random under-sampling technique can discard useful data.

2.13.2.2   Synthetic Minority Oversampling Technique (SMOTE)

Synthetic Minority Oversampling Technique (SMOTE) is used to avoid the over-fitting problem, which was proposed by Chawla *et al.* (2002). This technique is considered to be state-of-art and works well in many applications. Thus it generates synthetic data based on the feature space similarities for minority instances. The methodology used to create a synthetic instance is as follows: find the K-nearest neighbors for each minority instance by selecting one random instance, and then calculate its linear interpolations to produce a new minority instance in the neighborhood.

The synthetic minority oversampling technique (SMOTE) is a powerful and widely used method (Manohar, 2011) used to create artificial data based on feature space as a random set of minority class observations. It is mainly used to shift the classifier bias for the minority class and generate artificial data using bootstrapping and the k-nearest neighbor's algorithm. This can be achieved by:

1. Calculating the difference between the feature vector and its nearest neighbor.
2. Multiplying this difference between the feature vector and its nearest neighbor using a random number between 0 and 1.
3. Adding the multiplication result to the feature vector under consideration. This causes the selection of a random point along the line segment between two specific features.

2.13.2.3   ADASYN: Adaptive Synthetic Sampling

Motivated by SMOTE, He *et al.* (2008) invented the Adaptive Synthetic sampling (ADASYN) technique, which received wide attention. Their methodology depends on generating samples of the minority class according to their density distributions. More synthetic data is generated for minority class samples that are harder to learn, compared to those minority samples that are easier to learn by calculating the K-nearest neighbors for minority instance, and merging the class ratio of the minority and majority instances to generate new samples. By repeating this process, it adaptively shifts the decision boundary to focus on those samples that are difficult to learn.

2.13.3 COST-SENSITIVE LEARNING (CSL)

Cost-Sensitive Learning is a type of learning in data mining used to solve the classification problems with imbalanced data by evaluating the costs associated with misclassifying observations and taking these costs into consideration. The main difference between cost-sensitive learning and cost-insensitive learning is that cost-sensitive learning treats the different misclassifications in different ways. Cost-insensitive learning does not take the misclassification costs into consideration; it is used to obtain a high accuracy of classifying results compared to known classes. Treating the imbalanced data sets as they occur in many real-world applications, where the class distributions of data are highly imbalanced, can use cost-sensitive learning to solve this problem (Charles *et al.*, 2008).

This method highlights the imbalanced learning problem by using cost matrices which describe the cost for a misclassification scenario but do not create balanced data distribution. Thus, research has shown that cost-sensitive learning has outperformed sampling methods many times over; this method provides a useful alternative to sampling methods. (www.analyticsvidhya.com/blog/2016/03/practical-guide-deal-imbalanced-classification-problems/).

2.14 THEORETICAL CONCLUSION

Even though glaucoma progresses slowly, there is no reason to delay the patients' treatment of the disease as it can cause blindness. Early glaucoma detection is essential to manage the irreversible damage done to vision. Furthermore, glaucoma comes in various forms and is highly complicated, but the availability and development of various eye imaging instruments and CAD systems can now facilitate the identification of the glaucomatous structural changes, and provide patients with an early diagnosis. From the studies that have been discussed, a CAD system based on fundus images could help save millions of people from vision loss. By adapting therapy procedures to these new findings, it is possible to diagnose glaucoma in a fast, cheap, and accurate way, fulfilling the aims of this research.

3 Related Works

3.1 INTRODUCTION

Computer-aided Diagnosis (CAD), which can automate the detection process for glaucoma, has attracted extensive attention from clinicians and researchers too. It not only alleviates the burden on clinicians by providing objective opinions with valuable insights, but also offers early detection and easy access for patients, potentially saving their sight.

This chapter will present a review of the influential and exciting work related to this research, based on the glaucoma risk factors mentioned in (Chapter 2), the optic disc and optic cup segmentation methods, as well as the existing computerized approaches in the literature.

3.2 OD AND OC SEGMENTATION

This section will present an overview of several methods for OD and OC segmentation that have been evaluated by their authors on publicly available datasets, with both images and ground truths provided.

Juan *et al.* (2019) studied the application of various Convolutional Neural Networks (CNN) to compare the performance of relevant factors, such as the data set size, the architecture, and the use of transfer learning against newly defined architectures. They also compared the performance of the CNN-based system to human evaluators and evaluated the integration of images and data collected from the patients' clinical history. The best performance was in a transfer learning scheme with VGG19, which achieving an AUC of 0.94 and sensitivity and specificity ratios similar to the expert evaluators of the study. The results obtained using 3 databases with 2,313 images indicated that this solution could be an accurate option for the design of a Computer-Aided System for the detection of glaucoma in large-scale screening programs.

Huazhu *et al.* (2018) used a deep learning method named M-Net, which jointly used OD and OC segmentation in a one-stage, multi-label system. It comprised a multi-scale input layer, a U-shape convolutional network, side-output layer, and multi-label loss function. For improving the segmentation performance further, they also introduced polar transformation, which provided a representation of the original image in the polar coordinate system; the proposed method also obtained glaucoma detection with a calculated CDR value on both the ORIGA and SCES datasets.

Artem (2017) presented a universal approach for automatic optic disc and cup segmentation, which was based on deep learning, namely a modification of the U-Net convolutional neural network. Their experiments included comparison with the best-known methods on publicly available databases DRIONS-DB, RIM-ONE (version three), and DRISHTI-GS. The results showed a 96% dice coefficient for disc segmentation in the RIM-ONE database, and a dice coefficient of85% for disc segmentation in the DRISHTI-GS database.

Vaishnavi Kamat *et al.* (2017) presented a method to detect Glaucoma using the Enhanced K-Strange Points Clustering algorithm to obtain cup, disc, and blood vessels. The elliptical fitting method was used to compute CDR and ISNT, which were used as inputs to the Naïve Bayes classifier.

Kartik *et al.* (2017) detected glaucoma from Retinal Fundus Images by analyzing ISNT measurements and features of the optic cup and blood vessels algorithm, using four different features of ONH used to detect glaucoma. Their algorithm worked effectively because it used four features of ONH together and used the results in grouping the images into glaucomatous, healthy, or unresolved sets. Other research can be done by analyzing the textural changes of RNFL.

Almazroa *et al.* (2017) presented a novel OD segmentation algorithm based on applying a level-set method on a localized OD image, and preventing the blood vessels from interfering with the level-set process, an in painting technique was applied. The accuracy of the algorithm in marking the optic disc area and centroid was 83.9%. The results of the algorithm has been tested using 379 images.

As Maureen at (2017) assess the impact of manual and default threshold selection on the reliability and accuracy of skull STL models using different CT technologies tested using female and male human cadaver head were imaged by multi-detector row CT, dual-energy CT, and two cone-beam CT scanners. Four medical engineers manually thresholded the bony structures on all CT images.

The skull STL models generated by the lowest and highest selected mean threshold values, and the geometric variations between all manually thresholded were calculated. Furthermore, in order to calculate the accuracy of the manually and default thresholded STL models, all STL models were superimposed on an optical scan of the dry female and male skulls ("gold standard"). They found intra- and inter-observer variability of the manual threshold selection was good (intra-class correlation coefficients >0.9).

Jen *et al.* (2017) developed and trained a convolutional neural network to automatically segment optic disc, fovea and blood vessels; on average, segmentation correctly classified 92.68% of the ground truths (on the testing set from Drive database), where the highest accuracy achieved on a single image was 94.54% and the lowest was 88.85%. They found that a convolutional neural network can be used to segment blood vessels, and also optic disc and fovea with high accuracy.

In Sirshad *et al.* (2016) research, the disc segmented using of gradient extracted from line profiles that pass through optic disc margin. The evaluation results of OD segmentation achieved by jaccard coefficient, dice coefficient, and distance between OD centers are and obtained 0.8331, 0.9078, and 6.44, respectively.

Abdullah *et al.* (2016) presented a robust methodology for optic disc detection and boundary segmentation, which is a preliminary step in the development of CAD systems for glaucoma in retinal images. The proposed method is based on morphological operations, circular Hough transform and the grow-cut algorithm. The method is quantitatively evaluated on five publicly available retinal image databases: DRIVE, DIARETDB1, CHASE_DB1, DRIONS-DB, Messidor and one local Shifa Hospital Database and achieves an optic disc segmentation success rate of 100% for the above databases with 99.09% and 99.25% accuracy for the DRIONS-DB, and ONHSD databases, respectively. The OD boundary detection results obtained an average spatial overlap of 78.6%, 85.12%, 83.23%, 85.1%, 87.93%, 80.1%, and 86.1%, respectively, for these databases.

Prasad *et al.* (2015) presented optic disc and optic cup segmentation based on superpixel classification, before that adaptive histogram equalization and Gabor filter are used for classifying each superpixel as disc or non-disc and cup-to-disc ratio (CDR) is evaluated and compared with threshold value for detection of glaucoma.

Satish *et al.* (2015) propose segmentation of optic disc and optic cup using superpixel classification, and for optic disc segmentation, clustering algorithms are used to classify superpixels as a disc or non-disc. But for optic cup segmentation has been done by the Gabor filter and Thresholding added to the clustering algorithms.

Medha *et al.* (2014) investigated and compared the performance of five methods used for optic disc segmentation. These five methods were based on the use of the following algorithms: distance regularized level set, Otsu Thresholding, region growing, particle Swarm optimization and generalized regression neural network, all of which were tested on a single database. The method using the generalized regression neural network was found to be the best-suited application due to highest region agreement, lowest non-overlap ratio, lowest relative absolute area difference and low execution time.

For the optic cup segmentation task, authors Ingle and Mishra (2013) used a 2-layer multi-scale convolutional neural network trained with boosting. The training process pipeline was multi-staged and included patches preparation and neural network training. For pre-processing and entropy filtering, L*a*b* color space was performed to extract the most important points of an image, followed by the contrast normalization and standardization of patches. The Gentle AdaBoost algorithm was then used to train convolutional filters, which are represented as linear repressors for small patches. At the time of the test, image propagation through the network was followed by an unsupervised graph cut. The method was then evaluated on the DRISHTI-GS database, and found to outperform all other existing methods in terms of Intersection-over-Union score and Dice score. However, it is necessary to note that this method cropped images by area of their optic disc (cup) before performing segmentation of the optic disc (cup). This makes the method inapplicable to new, unseen images of full eye fundus, since it requires a bounding box of optic disc and cup to be available in advance.

Ingle and Mishra (2013) discussed cup segmentation based on the gradient method, as the gradient is the variation in the intensity or color of an image. The gradient images convolved with a filter. Two methods were used to find the gradient:

(1) Linear gradient.
(2) Radial gradient. The contrast was improved for all image components (red, blue, and green) by Contrast Limited Adaptive Histogram Equalization (Zuiderveld, 1994).

The initial threshold was set for red (R), blue (B), and G (green) components after much iteration to detect the region where R was less than 60, and B and G were more than 100, where other pixels were eliminated by setting their values to zero. The intensities were computed and linearly transformed to the range of (0–1). They found that G and B channels were more effective for OC segmentation. The circular structural elements were used to fill the blood vessels region in order to obtain a continuous region. The algorithm was evaluated based on the accuracy of the cup and disc area in all directions as well as CDR, instead of relying on the accuracy only in one direction. The algorithm can be extended to distinguish between the glaucomatous and normal images.

Jun *et al.* (2013) proposed optic disc and optic cup segmentation using superpixel classification for glaucoma detection. In optic disc segmentation they used histograms, and centered surround statistics to classify each superpixel as disc or nondisc. For optic cup segmentation, they used the histograms, center surround statistics, and the location information used as feature space to improve the performance. The proposed segmentation methods have been tested on 650 images with optic disc and optic cup boundaries manually marked as a ground truth. Experimental results obtained an overlapping error of 9.5% and 24.1% in optic disc and optic cup segmentation, respectively. The segmented optic disc and optic cup are then used to compute the cup-to-disc ratio for glaucoma detection, and achieved areas under curve of 0.800 and 0.822 in two data sets. Thus, this method can be used for segmentation and glaucoma detection, and the self-assessment can be used as an indicator of cases with large errors, enhancing the clinical deployment of automatic segmentation and detection.

Lalonde *et al.* (2001) proposed an OD localization scheme using a Hausdorff-based template matching and pyramidal decomposition. A similar method is proposed in (Pallawala *et al.*, 2004) with an improved morphological-based pre-processing step.

From the above studies, it can be concluded that OD is a challenging task mainly due to blood vessel occlusions, ill-defined boundaries, and image variations near disc boundaries due to pathological changes and variable imaging conditions. Specifically, the occurrence of similar characteristics/regions (atrophy) near the disc boundary, irregular disc shape and boundary are the most essential aspects to be addressed by an OD segmentation method, (see Chapter 2). Due to the high density of blood vessels in the OC, segmentation in this region is more difficult than OD segmentation. Furthermore, the gradual intensity change between the cup and neuroretinal rim causes extra complications for cup segmentation. In addition, glaucoma changes the shape of the cup region. Therefore, for these difficulties there is a need to find and modify new segmentation techniques to get high segmentation accuracy; in this research a modified thresholding segmentation method for OD segmentation was applied and the vessel was removed to overcome these difficulties.

3.3   DISC-TO-CUP RATIO (CDR)

This section will present a detailed review of the OD and OC segmentation methodologies that automatically detect OD and OC boundaries. These segmentation techniques help doctors to diagnose and monitor glaucoma by providing them with clear and accurate information regarding the ONH structure, allowing them to detect glaucoma based on CDR.

Sharanagouda *et al.* (2017) proposed a method unlike past works which relies on a single color channel for extracting the Optic Disk (OD) and Optic Cup (OC) used in CDR calculation, they proposed a novel combined color channel and ISNT rule-based automated glaucoma detection system, and found that the proposed method betters the single channel-based method, giving an overall efficiency of 97%.

Ranjith *et al.* (2015). Used technique used here is a core component of ARGALI (Automatic cup-to-disc Ratio measurement system for Glaucoma detection and Analysis), a system for automated glaucoma risk assessment. The algorithm's effectiveness was tested on segmented retina fundus images, by comparing the automatic cup height using ARGALI with the ground truth. The result was that the algorithm accurately detected neuro-retinal cup height. This work will improve the efficiency of glaucoma detection using fundus images of the eye.

In Prasad *et al.* (2015) research, the optic cup and optic disc segmentation was used for assessment of the optic nerve head. Segmentation was based on superpixel classification for both optic disc and optic cup segmentation. Firstly, for optic disc and optic cup segmentation, adaptive histogram equalization and Gabor filter were applied basically for classifying each superpixel as disc or non-disc. Finally, CDR was calculated and compared with the threshold value for the detection of glaucoma.

Kavitha *et al.* (2014) described a process to automatically locate the optic nerve in a retinal image. The optic nerve is one of the most important organs in the human retina and can be used to locate the OD position in fundus images, which is important for many reasons (such as for the detection of glaucoma). The method is based on the detection of the main retinal vessels; thus, the retinal vessels start from the OD, and their path follows a similar directional pattern in all images. Glaucoma detection is possible by basically using the medical history, intra-ocular pressure, and visual field loss tests of a patient, together with a manual assessment of the OD through ophthalmic testing. Since enlargement of the cup with respect to OD is an important indicator of glaucoma (called CDR), various parameters are estimated and recorded to assess the stage of glaucoma. Compared with previous methods, the results obtained smaller CDR errors and higher AUC in glaucoma detection by the proposed method. The proposed disc and cup segmentation was tested at the SCES dataset and achieved an AUC of 0.800, which is 0.039 lower than the AUC of 0.839 of the manual CDR computed from manual disc and manual cup, and a glaucoma detection result AUC 0.822, which is much higher than 0.660 by the currently used IOP measurement. They found the accuracy to be good enough for a large-scale glaucoma detection, but it is important to know how different partitions affect the performance.

Mahalakshmi (2014) proposed a method to segment the optic disc and optic cup using the Simple Linear Iterative Clustering (SLIC) algorithm, and K-Means clustering for glaucoma detection to obtain accurate boundary delineation. In optic disc and cup segmentation, histograms and center surround statistics are used to classify each super pixel as disc or non-disc, and the location information is also included into the feature space to boost the performance. The segmented optic disc and optic cup are then used to compute the CDR to confirm glaucoma for a patient.

Eleesa Jacob (2014) proposed an optic disc and optic cup segmentation was used for diagnosing glaucoma. The optic disc and optic cup segmentation was achieved using the super pixel classification technique. In optic disc segmentation, histograms and center surround statistics were used, and the quality of the optic disc segmentation was achieved using the self-assessment method. For optic cup segmentation, the location information was also included into the feature space to improve the performance in addition to the histograms and center surround statistics. Then segmented optic cup and optic disc was then used to compute the CDR to identify whether the given fundus image was glaucomatous or not.

The segmentation can be analyzed using the MATLAB®, as SobiaNaz *et al.* (2014) were to calculate the CDR automatically. Pre-processing methods such as anisotropic filtering have been performed, and automatic disc extraction was done using 3 techniques: 1) edge detection method, 2) optimal thresholding method, and 3) manual threshold analysis. The threshold level-set method was used for the cup segmentation and tested on the DRIVE database.

Cheng (2013) proposed the optic disc and optic cup segmentation for glaucoma detection. Using this method the optic disc segmentation, histograms, and center surround statistics are used for OD and OC segmentation, then calculating the CDR for glaucoma detection. This method's limitation was that it had poor visual quality. Morphological operations were used to improve the algorithm and locate the optic disc in retinal images as used by Angel Suero *et al.* (2013).

Achanta (2012) proposed to improve the segmentation performance by using a super pixel algorithm, the simple linear iterative clustering (SLIC) method. It empirically compares five state-of-the-art super pixel algorithms for their ability to adhere to image boundaries, speed, memory efficiency, and their impact on segmentation performance. However this method increased the computations.

Joshi *et al.* (2011) proposed an approach for an automatic OD parameterization technique based on segmented disc and cup regions obtained from monocular retinal images. The OD was segmented by using an active contour model, and by enhancing the Chan-Vese (C-V) model by including image information at the support domain around each contour point. The limitation of this method was that it did not provide better quality.

Babu (2011) proposed an algorithm for the measurement of CDR. Thus, it is considered as a parameter for the diagnosis of glaucoma. The result achieved 90% accuracy.

Muramatsu *et al.* (2011) again extended their previous work by presenting a technique for detecting the peripapillary atrophy (PPA), which is the other peripapillary feature of glaucoma. The objective of their procedure was to detect the PPA by using a texture analysis based on the gray-level co-occurrence matrix. In a dataset of 26 images, the sensitivity and specificity for detecting moderate to severe PPA regions was 73% and 95%, respectively.

Wong *et al.* (2010) developed an SVM-based model optic cup method for glaucoma detection using the CDR in retinal fundus images. Joshi G. D. *et al.* (2010) developed vessel bend-based cup segmentation in retinal images, and Shijian Lu *et al.* (2010) proposed a background elimination method for the automatic detection of OD.

Gopal *et al.* (2010) developed a deformable model guided by regional statistics to detect the OD boundary and OC boundary based on lab color space and the expected cup symmetry. This method uses sector wise information and gives rise to fewer false positives and hence better specificity. In the final results, the computed error value was less for a normal image than for a glaucomatous image.

Zhuo Zhang *et al.* (2009) designed a convex, hull-based, neuro-retinal, optic cup ellipse optimization technique. Hussain (2008) proposed a method for optic nerve head segmentation using genetic active contours, and Huajun *et al.* (2007) designed a fractal-based automatic localization and segmentation of optic disc in retinal images.

The Active Shape Model-based optical disc detection (ASM) was implemented by Huiqi *et al.* (2003). The parameters for this model were selected using the Principal Component Analysis technique. The faster convergence rate and the robustness of this technique were proved by experimental results.

A comprehensive review was provided by the algorithms used for OD and OC detection that lead to a diagnosis of glaucoma by CDR. Many algorithms were limited due to the complexities of ONH structure that appear in segmentation results, which are very variable among people and among imaging devices, and are affected by papillary atrophy and disc drusen, causing some difficulties for disc segmentation due to its similarity in intensity to disc boundaries, but there are some algorithms capable of segmenting the optic disc with PPA perfectly. On the other hand, disc drusen causes greater difficulty for segmenting since it covers boundaries, especially in advanced cases.

Other difficulties are blood vessels, which affect localizing OD and OC, and play an important role in the accurate segmentation of cup boundaries and represent a challenge facing many researchers. The algorithms performed differently depending on the datasets of images. Some approaches used a small dataset, while some used large datasets to train and test the algorithm. Also, the severity of the disease was different among the datasets used in different techniques; therefore, the corresponding algorithms cannot be compared with each other. Most of the OD segmentation was based on the circular Hough transform technique, along with other detection techniques.

In conclusion, segmentation of the optic disc and optic cup has been very important because it forms the basics of glaucoma diagnosis, although there are still chances for improvement in segmentation techniques. Only a few of the existing methodologies, whether for optic disc or for optic cup segmentation, can be applied for glaucomatous retinal images. Also, most of the current methods have been tested on a limited number of datasets, which do not provide images with many different characteristics. Furthermore, the generally low resolution of the images has made the segmentation process even more challenging. This can be handled by advanced cameras capable of taking high volumes of high-resolution retinal images. In order to achieve good outcomes for the images captured by different systems, robust and fast segmentation methods are required. Most of the retinal images used to evaluate segmentation methods have been taken from adults.

The glaucoma screening system complements, but does not replace, the work of ophthalmologists and optometrists in diagnosis; routine examinations have to be conducted in addition to the fundus image analysis. However, the system facilitates diagnosis by calculating the disc and cup structural parameters and showing greater details of ONH, such as the disc and cup areas, the vertical and horizontal cup-to-disc ratios, and cup-to-disc area ratio, and also checking the ISNT arrangement. This is a shareable opinion that could associate the worlds of consultant ophthalmologists, optometrists, orthoptists, and engineers.

The research contributions in optic disc and cup segmentations use simple, fast, inexpensive technique for segmentation (global thresholding), and can be modified to solve PPA and drusen problems in disc segmentation; then, tests in two databases are carried out to provide many different characteristics and achieve good outcomes, and the main contributions are the features, obtained from the OD and OC parts, which are color and shape features, considered to be new features rather than the common CDR, ISNT, and rim area mentioned in previous studies.

3.4 RNFL REVIEW

Oh *et al.* (2015) proposed an automatic detection method for RNFL defects on digital fundus images to detect and separate glaucomatous from non-glaucomatous cases. The RNFL defects detected here were based on the vertical dark bands, firstly the non-uniform illumination of the fundus image was corrected; secondly, the blood vessels were removed, and then images were converted to polar coordinates using the center of the optic disc. Finally, false positives (FPs) were reduced by using knowledge-based rules. The results obtained had a sensitivity of 90% and a 0.67 FP rate per image for the 98 fundus images with 140 RNFL defects and 100 fundus images of healthy subjects. The RNFL defects had variable depths and widths, with uniformly high detection rates, regardless of the angular widths of the RNFL defects. The overall accuracy was 0.94, 86% sensitivity, and 75% specificity.

The proposed CAD system successfully detected RNFL defects in digital fundus images. Thus, the proposed algorithm is useful for the detection of glaucoma. Imran Qureshi (2015) introduced a survey paper and various image-processing techniques, as well as different computer-based systems involved particularly in the detection and diagnosis of glaucoma, were discussed in detail. This paper was to highlight the severity of glaucoma across the world, and survey the research work done so far on this disease. The future directions regarding detection of glaucoma can be an evaluation of various algorithms discussed in this paper by implementing and testing them on a large amount of data. Similarly, various arguments like the neuro-retinal rim area, width, and vertical CDR can be calculated which indicate the development of glaucoma. Thus, intensity of glaucoma can be determined by using 3-D reconstruction image. Then the execution of optic cup segmentation approaches can be improved by using vessel observing and vessel in painting, as well as machine-learning methods for finding the best arguments in many patterns, such as threshold level set and edge detection. In future, more efforts are required for the improvement of classification methods and accuracy rates. There is a system required which accomplishes high execution by promoting a large number of data for making class and blending various detection approaches for the diagnosis of glaucoma.

Syed Hussain (2015) introduced a survey paper that depicts many works related to automated glaucoma detection, lowering eye pressure in glaucoma's symptoms based on early glaucoma diagnosis at early stages slows down the progression of the disease and helps save vision. Through the extensive literature review carried out it has been observed that there are various methods for detection of glaucoma with good results, but there is still a need to develop a Computer-Aided System which can not only help diagnose glaucoma at early stage, but would also help in checking the progression of the disease to be prevented. A lot of recent research is being carried out for the detection of glaucoma using fundus images, but still the detection of glaucoma's progression on inpatients remains to be researched. In the future, we need to develop more accurate, robust, and affordable automated techniques for glaucoma detection. Once glaucoma is diagnosed early and correctly, they can be treated and avoid total blindness.

In Oh J. E. *et al.* (2015), they proposed a fully automatic method for detecting various forms and widths of RNFL defects in color fundus images. Fundus photography is the most common screening tool to detect RNFL defects in various optic neuropathies. However, the detection of RNFL defects by using fundus photographs depends on the experience of the examiner, and early defects may be missed because of the low contrast of the RNFL. Therefore, they developed a simple and efficient algorithm to assist the ophthalmologist for the detection of RNFL defects. The strength of the proposed algorithm is that it can detect very narrow defects in early stage glaucoma, to non-glaucomatous optic neuropathy involving the papilla macular bundle accurately. Thus, no previous studies described specific methods for detecting RNFL defects with various forms and widths in fundus images. Their results showed that the proposed algorithm was successful, with a sensitivity of 90% for glaucoma and 100% for papilla macular bundle defects in non-glaucomatous optic neuropathies.

Siddeeqa *et al.* (2015) proposed a method to compare retinal nerve fiber layer (RNFL) thickness in black and Indian myopic students at the University of KwaZulu-Natal. A total of 80 (40 black and 40 Indian) participants of both genders, aged between 19 and 24 years (mean and standard deviation: 21 ± 1.7 years) were included in the study. Refractive errors were assessed with the Nidek AR-310A auto-refractor and via subjective refraction. RNFL thicknesses were then measured using the iVue-100 optical coherence tomography device.

Axial lengths were measured using the ultrasound device (Nidek US-500 A-scan). Data were analyzed by descriptive statistics, t-tests, Pearson's correlation coefficients, and regression analysis, and they found the mean myopic spherical equivalent was significantly more negative in the Indian people (-2.42 D ± 2.22 D) than in the black people (-1.48 D ± 1.13 D) (p = 0.02). The mean axial length was bigger in the black people (23.35 mm ± 0.74 mm) than in the Indian people (23.18 mm ± 0.87 mm), but the difference was not significant. The sample used (n = 80), and the average global RNFL thickness ranged from 87 μm to 123 μm (105 $\mu m \pm 9$ μm). Mean global RNFL thickness was slightly greater amongst black (108 $\mu m \pm 7$ μm) than amongst Indian (102 $\mu m \pm 9$ μm) (p = 0.00) participants. Mean global RNFL thickness was similar for male (106 $\mu m \pm 7$ μm) and female (105 $\mu m \pm 10$ μm) (p = 0.79) participants. A positive and significant association between myopic spherical equivalent and global RNFL thickness was found for the total sample (r=0.36, p=0.00) and for Indians (r=0.33, p=0.04), but not for the black (r=0.25, p=0.13) participants. The results obtained a negative and significant correlation between axial length and global RNFL thickness between the Indian participants (r=−0.34, p=0.03), but not between the total sample (r=−0.12, p=0.30) or the black (r=0.06, p=0.73) participants.

The findings suggest that racial differences in RNFL thickness need to be considered in the clinical examination and screening for glaucoma and other optic nerve pathologies amongst black and Indian people. Additionally, the possible influences of refractive error and axial length should be considered when evaluating RNFL thickness.

Jan Odstrcilika *et al.* (2014) here introduce a novel approach to capture these variations using computer-aided analysis of the RNFL textural appearance in standard and easily available color fundus images. The proposed algorithm built was based on the Gaussian Markov random fields and local binary patterns features, together with various regression models for prediction of the RNFL thickness. The algorithm described the changes in RNFL texture, by reflecting variations in the RNFL thickness. The method was tested on 16 healthy and 8 glaucomatous eyes. The results achieved significant correlation (normal s:=0.72±0.14; p-0.05, glaucomatous:=0.58±0.10; p 0.05) between the results of the predicted output and the RNFL thickness measured by optical coherence tomography, which is the standard glaucoma assessment device. The evaluation achieved good results to measure possible RNFL thinning.

In Iyyanarappan (2014) research a wavelet-based texture feature set has been used. The texture feature used was the energy of sub-images. Wavelet transform is a very efficient tool for feature extraction and it is very successfully used in biomedical image processing. Then the classification technique is developed to automatically detect the glaucoma from non-glaucoma cases. Texture Features by DWT achieved classification accuracy 95% and the system, easy to operate, non-invasive, and inexpensive. They have carried out the classification by Probabilistic Neural Network for the purpose of examining the efficiency of the features extracted. Future work, to use more powerful classifiers used, classification accuracy may further be improved.

Kumar (2014) introduced a review paper, and studied many works related to automated glaucoma detection. Glaucoma is one of the vital factors contributing to most of the blindness worldwide. So, there is a need to develop an inexpensive, automated technique for accurate diagnosis of different stages and types of glaucoma. These techniques will help in less developed countries where there is an acute shortage of ophthalmologists. In the future, there will be a need to develop more accurate, robust, and inexpensive automated techniques for glaucoma diagnosis, so that they can be useful to the poor. Thus, once glaucoma is correctly diagnosed, they can take proper medicine or undergo surgery in a timely manner to avoid total blindness.

Koprowski *et al.* (2014) proposed a new method for automatic determination of the RNFL (retinal nerve fiber layer) and other parameters, using mathematical morphology and profiled segmentation based on morphometric information of the eye fundus. A quantitative ratio of the quality of the optic disc and RNFL–BGA (bio morphological glaucoma advancement) was also proposed. The obtained results were compared with the results obtained from optic disc static perimetry. The result achieved a correlation with the static perimetry 0.78 for the existing method of image analysis, and 0.86 for the proposed method. The practical usefulness of the proposed ratio BGA and the impact of the three most important features on the result were assessed. The following results of correlation for the three proposed classes were obtained: cup/disc diameter 0.84, disc diameter 0.97, and the RNFL 1.0. Thus, analysis of the supposed visual field results in the case of glaucoma is possibly based only on OCT images of the eye fundus. The calculations and analyses performed with the proposed algorithm and BGA ratio confirm that it is possible to calculate the supposed mean defect (MD) of the visual field test based on OCT images of the eye fundus.

Pachiyappan *et al.* (2012) proposed a technique for Glaucoma diagnosis utilizing fundus images of the eye and the optical coherence tomography (OCT). The Retinal Nerve Fiber Layer (RNFL) can be classified into anterior boundary, which is the top layer of RNFL, the posterior boundaries (bottom layer of RNFL), and also the distance in between the two boundaries. glaucomatous and non-glaucomatous classification depends on the thickness of the nerve fiber layer, which is nearly 105 μm. This approach provided optical disc detection with a 97.75% accuracy.

Jan *et al.* (2012) proposed an automatic method to obtain RNFL texture analysis based on the combination of intensity, edge representation, and Fourier spectral analysis. DCFI with the size of 3504×2336 pixels (8 normal, 4 glaucomatous). The ability of proposed features to classify RNFL defects has been proven via comparison with OCT.

Prageeth *et al.* (2011) proposed an automatic method to obtain texture analysis using only intensity information about RNFL presence. DCFI with the size of 768×576 pixels (300 normal, 529 glaucomatous). Intensity criteria were used for detection of the substantial RNFL atrophy.

Glaucoma doctors consider signs like a patient's demographic data, medical history, vision measurement, and IOP (Intra Ocular Pressure), as well as the assessment from various types of imaging equipment. Following the clinical decision-making process, it is natural for us to design an automatic classifier that is able to combine inputs from multiple data sources. However, the limitations of the black-box manner of the supervised learning classifiers offers clinicians little insight as to how they work, thus hindering the deployment of such systems.

Acharya *et al.* (2011) carried out an automatic analysis of the RNFL texture using higher-order spectra, run length, and co-occurrence matrices. They used a DCFI with the size of 560×720 pixels (30 normal, 30 glaucomatous). The specificity of detecting glaucoma obtained an accuracy of over 91%. However, the article does not thoroughly explain how the features were extracted and in which area the image was computed.

Many algorithms have been proposed for detecting RNFL defects. Prageeth *et al.* (2011) used texture analysis by utilizing only intensity information about the RNFL around the optic disc in the red-free fundus image. Muramatsu *et al.* (2011) applied three sizes of Gabor filters to detect RNFL defects in the fundus image. The detection rates were 89% ~ 91% at 1.0 FPs per image. However, because determining the filter width for detecting RNFL defects with various forms and widths is difficult with these methods, they applied the Hough transformation to detect RNFL defect candidates with straight continuous lines. In addition, these previous studies confined their study subjects to those with glaucoma with localized RNFL defects, which would be apparently visible and found on OCT; however, these studies did not include early stage preparametric glaucoma, other non-glaucomatous optic neuropathies, and papilla macular bundle defects.

Odstrcilik *et al.* (2010) proposed the use of texture analysis by utilizing Gaussian Markov random fields (GMRF) for classification of healthy and glaucomatous RNFL tissue in fundus images. Their results were compared with the OCT images as a gold standard, using DCFI with a size of $3,504 \times 2,336$ pixels (18 normal, 10 glaucomatous). The feature's ability to differentiate between healthy and glaucomatous cases was validated using the OCT RNFL thickness measurement.

Bock *et al.* (2010) provided a competitive, reliable, and probabilistic glaucoma risk index (GRI) from images as its performance was comparable to medically relevant glaucoma parameters. This proves that data-driven methods are able to extract relevant glaucoma features. In future, it will be low-cost glaucoma detection methods like this that may help patients to do regular check.

The defects of the RNFL were most commonly located in the inferior temporal and superior temporal regions. These locations are the ones most frequently affected in the early stages of glaucoma. Additionally, the detection rate of the proposed algorithm was almost consistent, regardless of the angular widths of RNFL defects. However, among the total defects, the proposed algorithm performed worse in cases with shallow defects (i.e., in early stage glaucoma or in images with poor resolution). This study had several limitations. First, the number of images used in this study was not large; thus, a larger database should be used in the future. Second, they did not consider structural changes of the optic disc, such as cupping, notching, or the pallor of the rim.

The change in the optic disc is an important indicator of the severity of glaucoma, and the detection of optic disc parameters can provide a significant differential clue regarding glaucomatous and non-glaucomatous optic neuropathies. Therefore, future work should be focused on the detection of optic disc changes, and combining these findings with RNFL defects. Further, regarding the high rate of false positives per image, modification of the FP reduction method may improve the reliability of the current algorithm for the early detection of various optic neuropathies. In conclusion, the proposed algorithm showed a reliable diagnostic accuracy for automatically detecting RNFL defects in fundus photographs of various optic neuropathies.

Delia *et al.* (2006) showed that the segmentation of retinal layers from OCT images was fundamental in diagnosing the progress of glaucoma. This study has shown that the retinal layers can be automatically and/or interactively located with high accuracy with the aid of local coherence information of the retinal structure. OCT images are processed using the texture features analysis by means of the structure tensor with the complex diffusion filtering. Experimental results indicate that our proposed novel approach has good performance in speckle noise removal, enhancement, and RNFL lyre segmentation.

In conclusion, we have presented a survey of many works related to automated glaucoma detection see Table 3.1 and Table 3.2, where it has been observed through various methods that there is still a need to develop a Computer-Aided System that could not only help diagnose glaucoma, but would also help in checking the progression of the disease and prevent permanent blindness. Much of the recent research has used fundus images for the detection of glaucoma, but the detection of the disease's progression still needs to be researched further. For these reasons, there is a need to develop more accurate, robust, and affordable automated techniques for glaucoma detection.

TABLE 3.1
Comparison between Different Segmentation Methods

Author	Year	Segmented Part	Segmentation Method	Data Base	Results
Jun	2013	OD and OC	Superpixel	650 images	OD overlapping error 9.5% OC overlapping error 24.1%
Abdullah	2016	OD	Morphological Operation	4 Databases DRIVE DIARET_DB1 CHASE_DB1 DRIONS_DN	Spatial overlap 78.6% 85.1% 83.2% 85.1%
Sirshad	2017	OD	Line profile gradient	–	Jaccard 0.83 DSC 90%
Jen	2017	OD	Convolutional neural network	DRIVE	Accuracy 92.7%
Maureen	2017	Skull CT images	Manual threshold level	–	Correlation coefficient >.9
Almazroa	2017	OD	Level set	379 images	Accuracy 92.7%
Artem	2017	OD and OC	Deep learning U_NET	–	OD (DSC 96%) OC (DSC 85%)
Proposed Method	2018	OD and OC RNFL	Modified thresholding by circle reconstruction	RIM_ONE v1. RIM_ONEv2. DRISHTI_GS DB database	**OD** (SSIM 83%, DSC 90%, Jaccard 82%) **OC** (SSIM 93%, DSC 73%, Jaccard 60%) Overall accuracy 97%

TABLE 3.2

Comparison between Different Feature Types Used to Detect Glaucoma From Digital Fundus Images

Authors	Year	Features
Proposed System	**2018**	**Color, Shape, and Texture Features**
R. Geetha Ramani	2017	Statistical features
Mohd Nasiruddin	2017	CDR, blood vessel ratio, disc to center distance
Sharanagouda	2017	CDR + ISNT
Claro M.	2016	Disc segmentation, texture feature
Salem	2016	CDR, texture- and intensity-based features
Swapna	2016	Fractal Dimension + LBP
Oh, Yang	2015	RNFL defects
Abir	2015	Grid Color Moment
Morris T.	2015	BRIEF
Karthikeyan	2015	LBP + Daugman's algorithm
Iyyanarappan	2014	DWT
Eleesa Jacob	2014	CDR
Geetha Ramani	2014	Color spaces, channel extraction, statistical, histogram
Guerre, A.	2014	CDR
Jan	2014	RNFL
Maya	2014	Local Binary Patterns
Mahalakshmi	2014	CDR
Ganesh Babu	2014	CDR+ISNT
Kavitha	2014	CDR
Preethi	2014	CDR
Fauzia	2013	CDR, ISNT
Cheng	2013	CDR
Rama	2012	HOS, TT, and DWT
Achanta	2012	CDR
Pachiyappen	2012	RNFL
Babu	2011	CDR
Muramatsu	2011	PPA
Wong	2010	CDR

Compared with the related research mentioned in Table 3.1 and Table 3.2, this research contribution in OD segmentation is the new modification method based on global thresholding, reconstructing the segmented optic disc to be a full circle based on the truth that the optic disc is circular, which increases the segmentation accuracy and extracts the RNFL part in simple, accurate methods by subtracting the ROI image from the optic disc part, because the RNFL is the layer surrounding the disc part, and then using new color, shape, and texture features to detect glaucoma.

4 Research Methodology

4.1 INTRODUCTION

This chapter illustrates the materials and methods used in this research. The RIM-ONE database is considered to be the gold standard for ONH segmentation. Figure 4.1 shows normal and abnormal glaucomatous images taken from the RIM-ONE database. Even if the images are clear, one cannot distinguish between normal (healthy) images and abnormal (glaucomatous) images by just using the naked eye alone, which necessitates the development of an automated system.

4.2 DIGITAL FUNDUS IMAGE DATASETS

Fundus images are considered as raw materials to be enhanced, segmented, and evaluated. The image datasets are normally accompanied by a ground truth which acts as a benchmark for comparing and evaluating the achieved experimental results using the true results of that image set.

4.2.1 RIM-ONE DATABASE (VERSION TWO)

To assess the performance of the proposed glaucoma detection system, the digital retinal Image database for Optic Nerve Evaluation (RIM_ONE) database is used, containing a total of 158 fundus images: 118 images for normal cases and 40 images for glaucomatous cases. The camera used to capture these images is the Nidek AFC-210 fundus camera, mounted on the body of a 21.1 megapixel Canon EOS 5D Mark II. This database was designed as part of a research project developed in collaboration with three Spanish hospitals: Hospital Universitario de Canarias, Hospital Clínico San Carlos, and Hospital Universitario Miguel Servet. The main differences between all other databases and the RIM-ONE database are as follows: RIM-ONE is exclusively focused on ONH segmentation. It has a relatively large amount of high-resolution images (158), as well as manual reference segmentations for each image, created in collaboration with five glaucoma experts from the aforementioned hospitals (four ophthalmologists and one optometrist) to deliver reliable gold standards for decreasing the variability between experts' segmentations and the development of highly accurate segmentation algorithms (Fumero *et al.*, 2011). Figures 4.2 and 4.3 show the normal and abnormal fundus images in the database respectively.

4.2.2 DRISHTI-GS DATABASE

The DRISHTI-GS database is a comprehensive dataset of retinal images, which includes both normal and glaucomatous eyes, along with manual segmentations from multiple human experts. Both area and boundary-based evaluation measures

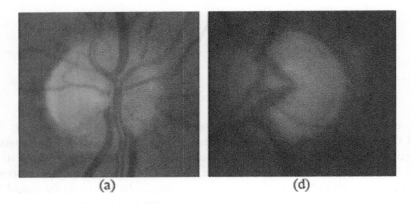

FIGURE 4.1 Shows RIM-ONE database images: (a) normal image, (d) glaucomatous image (abnormal), an example of the images used in the research.

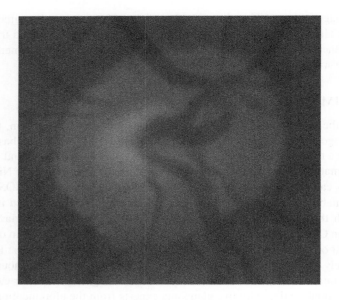

FIGURE 4.2 Shows examples of normal fundus images from the RIM-ONE database, where the cup is too small as indicator for healthy.

are presented to evaluate a method of various aspects relevant to the problem of glaucoma assessment, which contains 101 images, 70 glaucomatous and 31 healthy (Sivaswamy *et al.*, 2014). The optic cup segmentation methodology and final features are evaluated on the DRISHTI-GS database, as it is the only public dataset, to our knowledge, that consists of the manual demarcation of the optic disc and cup boundaries by clinical experts. Figures 4.4 and 4.5 show examples of the database images.

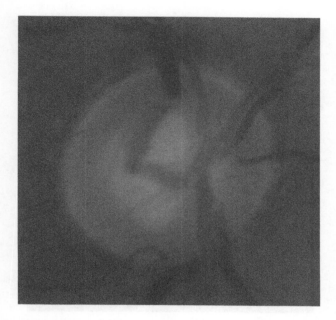

FIGURE 4.3 Shows examples of abnormal fundus images from the RIM-ONE database, where the cup enlargements are visible as an indicator for glaucoma.

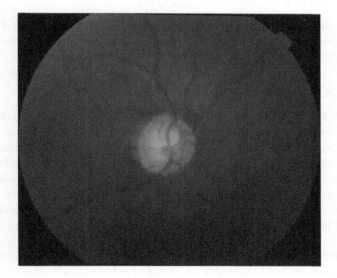

FIGURE 4.4 Shows an example of the original fundus images from the DRISHTI_GS database.

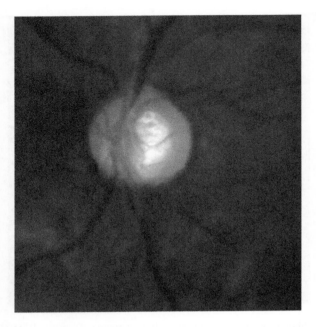

FIGURE 4.5   Shows examples of ROI fundus images from the DRISHTI_GS database after cropping the ONH part, which is affected by glaucoma.

4.2.3 RIM-ONE DATABASE (VERSION ONE)

This is the first version in the RIM-ONE database series, and contains 455 images (200 glaucomatous images and 255 suspected as potentially glaucomatous images) with the same ground truth mentioned in (Section 4.2.1).

4.3 PROGRAMMING ENVIRONMENT

MATLAB® is a high-level language used for visualization, numerical computation, and programming. It can be used in various applications, such as image and video processing, signal processing, control systems, bioinformatics, and communications. The simulation of the system proposed for glaucoma detection is done in MATLAB® R2016, http://cimss.ssec.wisc.edu/wxwise/class/aos340/spr00/whatismatlab.htm.

The proposed algorithm provides an automated glaucoma detection method based on a CAD system that helps ophthalmologists diagnose glaucoma early and with high accuracy. The algorithm takes a pre-processed digital fundus image and segments the optic cup, optic disc, and RNFL followed by a combined features (Arwa *et al.*, 2018) extraction from the segmented parts to train the classifier and test the system. The results are then used to classify the image as either glaucomatous or healthy.

4.4 METHODOLOGY

Figure 4.6 depicts the methodology used and clarifies the following steps.

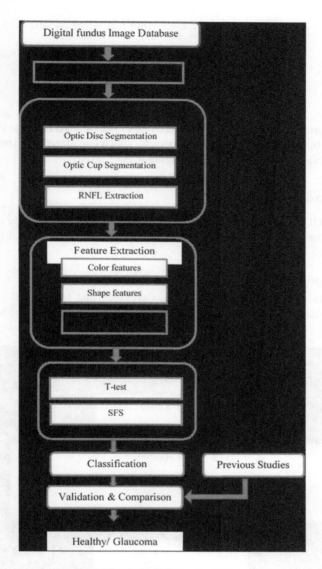

FIGURE 4.6 Proposed methodology used to diagnose glaucoma disease.

4.4.1 IMAGE PRE-PROCESSING AND ENHANCEMENT

The image pre-processing processes involved are: resizing the image to 256×256 pixels so that it has the specified number of rows and columns to reduce computational time, before moving on to the de-noising stage with median filters (5×5) (Arwa *et al.*, 2018), using the formula:

$$f(x,y) = \text{median}\{g(s,t)\}, s,t \in Sxy \qquad (4.1)$$

Then contrast enhancement is carried out using histogram equalization to enhance the image:

$$g_{i,j} = \text{floor}\left((L-1) \sum_{n=0}^{f_{i,j}} p_n \right) \tag{4.2}$$

Before any procedures are made, the red channel is extracted because it appears to have a good boundary and fewer blood vessels, which can affect segmentation accuracy by facilitating intensity analysis (Figure 4.7).

4.4.2 SEGMENTATION

The task of OD and OC segmentation is an important step to relatively quantify the changes in the OD and OC regions for evaluating glaucoma in the digital fundus image (Figure 4.8).

4.4.2.1 Optic Disc Segmentation

The optic disc segmentation algorithm is developed and tested in the RIM-ONE public database. The proposed method (Arwa *et al.*, 2018) can be divided into four

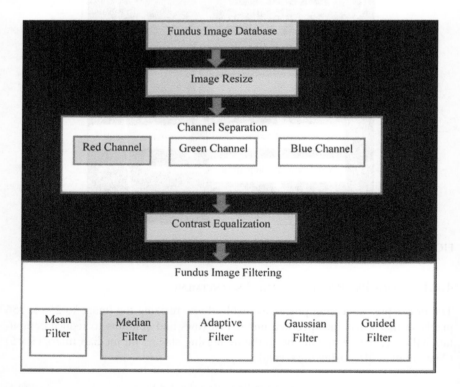

FIGURE 4.7 Shows the pre-processing steps, the gray box was the selected methods.

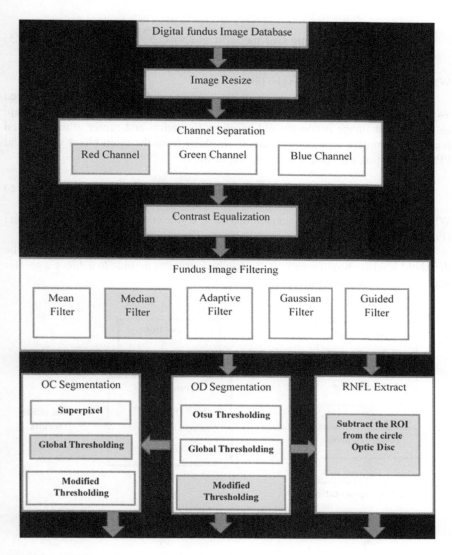

FIGURE 4.8 Shows the segmentation steps, the gray box was the selected methods.

steps. The **first step** is to remove the vessel in order to get an accurate segmentation, and this is done by morphological operation. Opening operation:

$$f \circ s = (f \ominus s) \oplus s \tag{4.3}$$

Closing operation: $f \bullet s = (f \oplus s) \ominus s$ (4.4)

The **second step** involves applying a thresholding level of 180 to segment the disc region from the red channel, which appears to best contrast with the disc region:

$$\text{Fixed thresholding is of the form: } g(x, y) = \sum \begin{array}{ll} 0 & f(x,y)<t \\ 1 & f(x,y) \geq t \end{array} \qquad (4.5)$$

Where t is the thresholding level $= 180$.

After that, the **third step** involves the boundary being smoothed and cleared away from unconnected objects. The binary image is obtained during the **final step**, as the circular image is constructed based on the radius and center of the detected region to minimize segmentation errors resulting from the main blood vessels and PPA surrounding the optic disc, using the formula:

$$(x - h)2 + (y - k)2 = r2 \qquad (4.6)$$

Where $r =$ the radius from segmented object, h, $k =$ the center from segmented object (Figure 4.8). The method's proposed steps are illustrated in Figure 4.9.

The vessel is removed via the opening morphological operation, and then the threshold level is chosen manually using a histogram as a guide to achieve a greater degree of accuracy in segmentation.

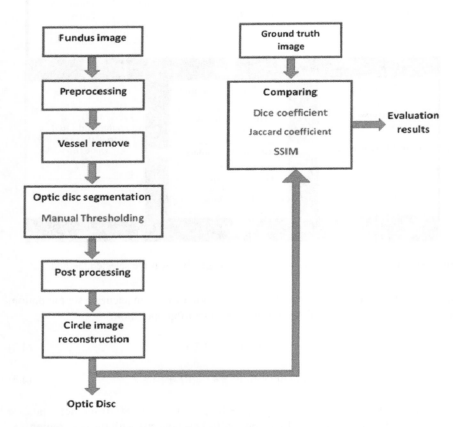

FIGURE 4.9 Proposed method for optic disc segmentation algorithm.

The post-processing step is done in order to smooth the boundary, carrying out the erosion and dilation operation mentioned in Chapter 2. This operation either expands or thickens the foreground objects in an image that are applied to the segmented disc, and the border is cleared to suppress structures which are lighter than their surroundings and connected to the disc border. For the segmented disc, the algorithm tends to reduce the overall intensity level in addition to suppressing the border, and a binarization of the image is needed to convert an intensity image into a binary image with a global threshold level, in order to obtain a more accurate segmentation. Based on the truth of the optic disc and optic cup, a logical circular image was constructed using the center point and the axis from the segmented object.

4.4.3 Optic Cup Segmentation

The enlargement of the cup region is an important indicator of the progression of glaucoma in the eye. The detection of the cup boundary from a retinal image is the most challenging task because depth is the best marker for the cup, which is lost in the two-dimensional projection. In absence of three-dimensional information, glaucoma experts use two-dimensional visual images to determine the cup boundary.

4.4.3.1 Cup Proposed Method

The optic cup segmentation method proposed by Arwa *et al.* (2018) depends on the optic disc thresholding level being set at 240, which is the best level for differentiating between the disc and cup. Then the border is cleared up, the boundary is smoothed, and the final image is binarized.

$$\text{Fixed thresholding is of the form: } OC(x,y) = \sum \begin{matrix} 0 & f(x,y) < t \\ 1 & f(x,y) \geq t \end{matrix} \qquad (4.7)$$

Where t is the thresholding level $= 240$, the proposed method illustrated in Figure 4.10.

The ROI was selected to be the optic disc because the optic cup is a circular, yellowish part inside the disc that can cause a minimal segmentation error to occur due to PPA, blood vessels, or the disc boundary itself. Thresholding is the most common method of segmenting images into particle regions and background regions. A typical processing procedure would start with filtering or other enhancements to sharpen the boundaries between objects and their background. Then, the objects would be separated from their background using thresholding. Here, the ROI was the entire optic disc, which was already filtered and had the blood vessel removed in the previous step (Section 4.4).

Basic global thresholding includes:

1) Selecting an initial estimate for T.
2) Segmenting the image using T. This will produce two groups of pixels. G1 consisting of all pixels with gray level values >T, and G2 consisting of pixels with values ≤T.

Where the threshold level (T = 240) can be chosen manually, or by using automated techniques, manual threshold level selection is normally done by trial and error, using a histogram as a guide.

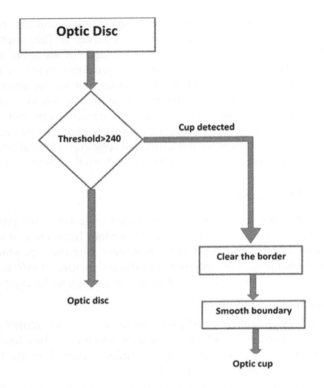

FIGURE 4.10 Cup segmentation proposed algorithm based on thresholding technique from optic disc region.

4.4.4 RNFL ROI Extraction

The RNFL (retinal nerve fiber layer) is the area surrounding the optic disc. For RNFL extraction a modified mathematical method has been used. This method is based on subtracting the whole image from the disc area. The goal of RNFL extraction is to measure this area thickness using the texture features, which will help in glaucoma detection.

$$RNFL = ROI\ region - OD \qquad (4.8)$$

4.4.5 Feature Extraction Step

Depending on an ophthalmologist's observation, glaucoma is diagnosed by an examination of the size, structure, shape, and color of the optic nerve head and RNFL surrounding the optic nerve itself. Based on this theory, a combination of texture, color, and shape features were applied to the disc, cup, and RNFL to classify glaucomatous cases from healthy cases. These features are:

1. Shape features, 13 features applied for OD and OC.
2. Color features, 3 features applied for OD, OC, and RNFL for each channel.
3. Texture features, 25 features applied for RNFL.

Shape features. In this research, the measurement of several properties describes the shape for each disc and cup within an image using the built-in "region prop" function in MATLAB®, which computes the following: *Area, Euler Number, Centroid, Filled Area, Filled Image, Extent, Major Axis Length, Minor Axis Length, Solidity, Equivalent diameter and Perimeter*; these measurements are applied for disc and cup masks. For measuring these features (Arwa *et al.*, 2018), a binary image is obtained. Some examples of the equations used are shown as follows (for more detail, see Section 2.7):

Centroid $(\vartheta x, \vartheta y)$ is:

$$\vartheta x = \frac{1}{N} \sum_{i}^{N} = 1xi \tag{4.9}$$

$$\vartheta y = \frac{1}{N} \sum_{i}^{N} = 1yi \tag{4.10}$$

N is the number of points in the shape.

$$\textbf{Solidity} = A_s / H \tag{4.11}$$

$$\textbf{Area} = (4 * \text{Area} / pi) \tag{4.12}$$

Color features. These are measurements that characterize the color distribution in an image in the same way that central moments uniquely describe a probability distribution. Color moments are mainly used for color indexing purposes as features in image retrieval applications in order to compare how similar two images are based on their color. Three color moments are computed per channel, and nine color features are computed for each disc, cup, and RNFL; these features are mean, standard deviation, and skewness.

Mean: the first color moment can be interpreted as the average color in the image.

$$Ei = \sum_{j}^{N} 1\frac{1}{N} P_{ij} \tag{4.13}$$

Standard Deviation: the second color moment is the standard deviation, which is obtained by taking the square root of the variance of the color distribution.

$$\sigma i = \sqrt{\left(\frac{1}{N} \sum_{j=1}^{N} \left(P_{ij} - E_i\right)^2\right)} \tag{4.14}$$

Skewness: the third color moment is the skewness. It measures how asymmetric the color distribution is, and thus it gives information about the shape of the color distribution (see Section 2.8).

$$S_i = \sqrt[3]{\left(\frac{1}{N} \sum_{j=1}^{N} \left(p_{ij} - E_i \right)^3 \right)} \tag{4.15}$$

Texture features. Many methods can be used to describe the main features of the textures, such as directionality, smoothness, coarseness, and regularity. Gray-level co-occurrence matrices are among the most important measurements used to describe texture; in this research, two techniques are used to describe the RNFL.

GLCM method: This is a method used to calculate the spatial relationship of pixels. The gray-level matrix, also known as the gray-level spatial dependence matrix, characterizes the texture of an image by calculating how often pairs of pixels with specific values that are also in specific spatial relationships occur in an image, thus creating a GLCM, and then extracting the following statistical measures:

Autocorrelation, Contrast, Correlation, Cluster prominence, Homogeneity, Cluster shade, Difference variance, Dissimilarity, Energy, Entropy, Maximum probability, Sum of squares, Sum average, Sum variance, Sum entropy, Difference entropy, Information measure of correlation, Inverse difference, Inverse difference normalized, and Inverse difference moment normalized. Below are some examples of GLCM features equations (see Section 2.10.1).

Energy feature

$$\text{Energy} = \sum_{i,j=0}^{N-1} \left(p_{ij} \right)^2$$

Entropy feature

$$\text{Entropy} = \sum_{i,j=0}^{N-1} -\ln\left(p_{ij} \right)\left(p_{ij} \right)$$

Contrast feature

$$\text{contrast} = \sum_{i,j=0}^{N-1} p_{ij} \left(i - j \right)^2$$

Homogeneity feature

$$\text{Homgeneity} = \sum_{i,j=0}^{N-1} \frac{p_{ij}}{1\left(i - j2 \right)}$$

Correlation feature

$$\text{correlation} = \sum_{i,j=0}^{N-1} p_{ij} \frac{\left(i - \mu \right)\left(j - \mu \right)}{\sigma^2}$$

Shade feature

$$\text{Shade} = \text{sgn}(A)A \, {}^{1}\!/_{3}$$

Prominence feature

$$\text{Prominence} = \text{sgn}(B)B \, {}^{1}\!/_{4} \tag{4.16}$$

Tamura method: a method used to calculate Coarseness, Contrast, and Directionality features for a digital fundus image:

Coarseness: is the most fundamental texture feature which, since it has a direct relationship to scale and repetition rates, aims to identify the largest size at which a texture exists, even a smaller micro texture.

$$A_k(x,y) = \sum_{i=x-2^{k-1}-1}^{x+2^{k-1}-1} \sum_{j=y2^{k-1}}^{y+2^{k-1}-1} f(i,j)/2^{2k} \qquad (4.17)$$

Contrast: is a statistical distribution of the pixel intensity obtained.

$$F_{\text{con}} = \frac{\sigma}{\alpha_4^{1/4}}$$

Direction: degrees we need to calculate the direction of the gradient vector, calculated at each pixel (see Section 2.11).

$$F_{\text{dir}} = \sum_{p}^{np} \sum_{\emptyset \in \omega\rho} (\theta - \theta\rho) 2H_D(\emptyset) \qquad (4.18)$$

4.4.6 FEATURES SELECTION

This is the process of selecting a subset of relevant features (variables, predictors) to be used in model construction. In this research, two types of features were selected:

A. Sequential feature selection;
B. t-test feature selection.
   SFS: the removal (or addition) of one feature at a time, based on the classifier's performance, until a feature subset of the desired size is reached.
   t-test: the t-test is any statistical hypothesis in which the test's statistics follow a student's t-distribution under the null hypothesis. It is used to determine if two sets of data are significantly different from each other, and it does this by assessing whether the means of two groups are statistically different from each other. Here, 78 features were applied to the OD, OC, and RNFL. The relevant features were then chosen by these two selection methods, resulting in eight features that were used in the final glaucoma classification.

4.4.7 CLASSIFICATION

The selected features of the images represent that they were generated from feature selection, and were used to detect glaucoma using the aforementioned classification methods. In this research, a mix of SVM, KNN, and ensembles classifiers were applied to choose the one with the best accuracy.

4.4.8 VALIDATION AND COMPARISON

The proposed algorithm was evaluated on classifier accuracy, sensitivity, and specificity compared with the ground truth attached to the databases; the final validation of the system was based on classifier outputs, run times, and a comparison with previous studies at each step: filtering, segmentation, and classification via many parameters.

4.4.8.1 Evaluation Parameters

I. Filters' Evaluation Parameters

To evaluate the performance of these filters, four parameters were applied to the original image and the filtered image. These parameters are:

- **Mean-Squared Error (MSE)**

$$\text{MSE}(x, y) = \frac{1}{N} \sum_{i=1}^{N} (x_i - y_i)^2 \tag{4.19}$$

The error signal $ei = xi - yi$ is the difference between the original image and the distorted image.

- **Peak Signal-to-Noise Ratio (PSNR)**

$$\text{PSNR} = 10 \log_{10} \frac{L^2}{\text{MSE}} \tag{4.20}$$

Where L is the dynamic range of allowable pixel intensities. For example, for an 8-bit per pixel image, $L = 2^8 - 1 = 255$. L is the dynamic range of allowable image pixel intensities.

- **Structural Similarity Index (SSIM)**

$$\text{SSIM}(x, y) = \left[l(x, y) \right]^\alpha . \left[c(x, y) \right]^\beta . \left[s(x, y) \right]^\gamma \tag{4.21}$$

Where it is a combination of the luminance, contrast, and structure similarity functions.

$$\text{Luminance function: } l(x, y) = \frac{2_{\mu x \mu y} + c_1}{\mu 2_x + \mu 2_y + c_1} \tag{4.22}$$

$$\text{Contrast function: } c(x, y) = \frac{2\sigma_x \sigma_y + c_2}{\sigma_x^2 + \sigma_y^2 + c_2} \tag{4.23}$$

$$\text{Structure function: } s(x, y) = \frac{2\sigma_{xy} + c_3}{\sigma_x \sigma_y + c_3} \tag{4.24}$$

$$\text{Standard deviation: } \sigma_x = \left(\frac{1}{N-1} \sum_{i-1}^{N} (x_i - \mu_x)^2 \right)^{\frac{1}{2}} \tag{4.25}$$

$$\text{Mean intensity: } \mu_x = \bar{x} = \frac{1}{N} \sum_{i=1}^{N} x_i \tag{4.26}$$

- **Signal-to-Noise Ratio (SNR)**

$$\text{SNR} = \frac{P_s}{P_N} = \frac{(\text{Asignal})^2}{(\text{Anoise})^2} \tag{4.27}$$

Where *Asignal* is the signal amplitude and *Anoise* is the noise amplitude, (Peter *et al.*, 2011).

II. **Segmentation Evaluation Parameters**
- The Dice similarity coefficient (DSC) is used as a statistical validation metric to evaluate the segmentation and spatial overlap accuracy; A and B are target regions, and these are defined as DSC:

$$\text{DSC} = 2(A \cap B) / (A + B) \text{ where } \cap \text{ is the intersection} \tag{4.28}$$

- Jaccard similarity coefficient is the Intersection over Union, which is a statistic used for comparing the similarity and diversity of image sets with a range of 0%–100%. The higher the percentage, the more similarity there is in the two image sets. This formula is:

$$J(A,B) = |A \cap B| / |A \cap B| \tag{4.29}$$

- Structural Similarity (SSIM) Index. The SSIM metric is a combination of local image structure, luminance, and contrast in a single local quality metric. It is where structures represent the patterns of pixel intensities among neighboring pixels, after normalizing for luminance and contrast. The SSIM metric is too close to the human visual system because of how good the human eye is at perceiving structure.

$$\text{SSIM}(x,y) = \left[l(x,y)\right]^\alpha \cdot \left[c(x,y)\right]^\beta \cdot \left[s(x,y)\right]^\gamma \tag{4.30}$$

$$l(x,y) = \frac{2(1+R)}{1 + (1+R)^2 + \dfrac{C_1}{\mu_x^2}} \tag{4.31}$$

$$c(x,y) = \frac{2\sigma_x\sigma_y + C_2}{\sigma_x^2 + \sigma_y^2 + c_2} \tag{4.32}$$

$$s(x,y) = \frac{\sigma_{xy} + C_3}{\sigma_x\sigma_y + C_3} \tag{4.33}$$

Where μx, μy, σx, σy, and σxy are the local means, standard deviations, and cross-covariance for images x, y. If $\alpha = \beta = \gamma = 1$ (the default for Exponents) and $C_3 = C_2/2$ (default selection of C_3), then the index simplifies to:

$$\text{SSIM}(x,y) = \frac{(2\mu_x\mu_y + C_1)(2\sigma_{xy} + C_2)}{(\mu_x^2 + \mu_y^2 + C_1)(\sigma_x^2 + \sigma_y^2 + C_2)} \tag{4.34}$$

III. **Classification Evaluation Parameters**

Classification is evaluated by two parameters: confusion matrix and ROC curve.

1. Confusion matrix: a plot to understand how the selected classifier performed in each class. The confusion matrix helps to identify the areas where the classifier has performed poorly; the rows show the true class, and the columns show the predicted class. The diagonal cells show where the true class and predicted class match (TP, TN, FP, and FN). If these cells are green and display high percentages, the classifier has performed well and is classified, which is then used to calculate the classifier's accuracy, sensitivity, and specificity with the following equations:

$$\text{SENSITIVITY} = \frac{\text{TP}}{\left(\text{TP} + \text{FN}\right)} \tag{4.35}$$

$$\text{SPECIFITY} = \frac{\text{TN}}{\left(\text{FP} + \text{TN}\right)} \tag{4.36}$$

$$\text{ACCURACY} = \frac{\left(\text{TN} + \text{TP}\right)}{\left(\text{TN} + \text{TP} + \text{FN} + \text{FP}\right)} \tag{4.37}$$

2. ROC Curve: the receiver operating characteristic (ROC) shows the true positive rate (TPR) versus the false positive rate (FPR) for the currently selected classifier.

4.4.9 GRAPHICAL USER INTERFACE

After the algorithm was compared with the previous study the GUI was built to help doctors deal with the software, and see the final decision and diagnosing steps. A major advantage of the Glaucoma CAD GUI is that it makes diagnosis more intuitive, as well as easier to learn and use.

The graphical user interface shows the diagnosis decision supported with the image of the optic disc and optic cup that have been segmented apart, as well as the selected feature values.

5 Results and Discussion

5.1 INTRODUCTION

In this chapter the experimental results of the proposed glaucoma detection system are discussed and their evaluations are explained in three sections. The **first section** for pre-processing and segmentation, the **second section** for glaucoma detection, and the **third section** for glaucoma classification. The key characteristic features used in the research system are color, shape, and texture features, which are the quantitative measures for the diagnosis of glaucoma. The quantitative evaluation of the proposed system is performed using the computed features compared with the gold standard database diagnosis based on standard parameters discussed in Section 2.14. The performance of the OC and OD segmentation is analyzed using the thresholding algorithm. A comparison is also made using the Ensembles and SVM classification approaches. The performance of the glaucoma classification system is evaluated using the receiver operating characteristic and area under ROC curve (AUC), for it is the diagnostic ability of the binary classifier.

5.2 SECTION ONE: IMAGE PREPARING

5.2.1 IMAGE PRE-PROCESSING AND ENHANCEMENT STEP

The aim of image enhancement is to improve the interpretability or perception of information included in the image for human viewers and to provide better input for pattern recognition stage, as in Figures 5.1 and 5.2.

5.2.2 CHANNEL SEPARATION RESULTS

After the three channels were extracted (red, green, blue), the best channel was found to be the red channel because it appeared to have a good boundary and fewer blood vessels, which affects segmentation accuracy. It has been used by researchers before (Claro M. *et al.*, 2016; Thresiamma *et al.*, 2015). Figure 5.3 shows examples of RGB channels for healthy and glaucomatous images.

5.2.3 IMAGE NOISE FILTRATION RESULTS

To remove the noise from digital fundus images, such as Gaussian and impulse (salt and pepper) noise, five filters were applied to the whole database. The output was then compared based on four parameters (see Chapter 2) after experiments, and, based on previous studies, the median filter was found to have the best performance (see Section 2.4.6), then the median filter was applied to the final algorithm, as in Figure 5.4. The results are shown in Figure 5.5 and Figure 5.6 (see also Tables 5.1 and 5.2).

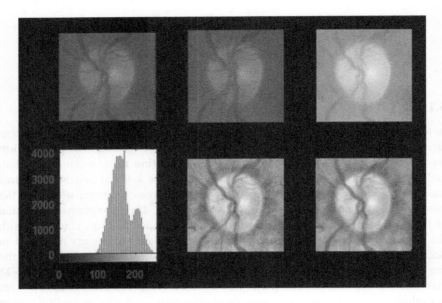

FIGURE 5.1   Shows the steps of an image enhancement for a healthy image. It also shows the original image, the image after it has been resized, the channel selected, the image histogram (x-axis to show intensity levels and y-axis intensity values), and the image after histogram equalization and filtering.

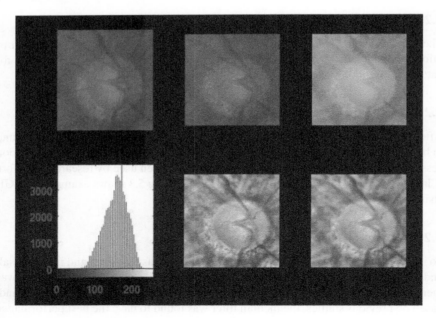

FIGURE 5.2   Shows the steps of image enhancement for a glaucomatous image. It also shows the original image, the image after it has been resized, the channel selected, the image histogram (x-axis to show intensity levels and y-axis intensity values), and the image after histogram equalization and filtering.

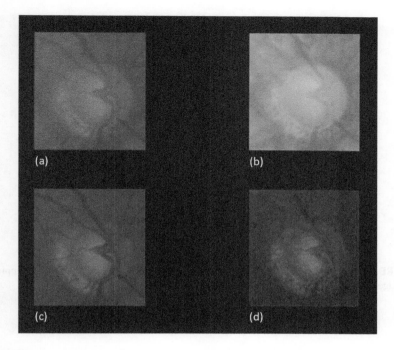

FIGURE 5.3   Shows digital fundus image color channel extraction as the red, green, and blue channels, in order to select the one that is best suited for segmentation.

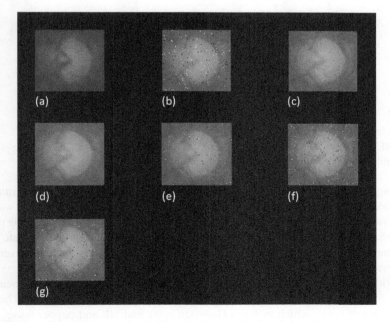

FIGURE 5.4   Shows an example of different filters applied in (salt and pepper) a noisy glaucomatous image: the original image, mean, median, adaptive, Gaussian, and guided filters.

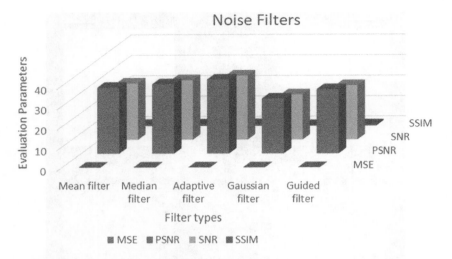

FIGURE 5.5   Bar chart showing the comparison between the different filter types applied to an image corrupted by Gaussian noise.

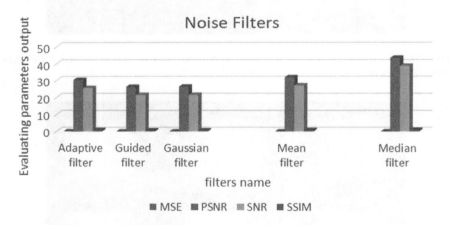

FIGURE 5.6   Bar chart showing the comparison between the different filter types applied to the original image corrupted by salt-and-pepper noise.

From the evaluation parameters noted above, the best performance against impulse noise was the median filter, and the best performance against Gaussian noise was the adaptive filter. To get more accurate results, the performance of the five filtering methods were tested with different noise levels, because the noise levels differentiate from one image to another and are affected by many factors, such as patient movement. The results are shown in Tables 5.3 and 5.4 and Figures 5.7 and 5.8.

Various filtering techniques such as linear filtering (Gaussian), non-linear filtering (mean, median), and adaptive median filtering have been applied. The two main types of noise, which includes Gaussian noise and salt-and-pepper noise, were tested, and the result was that the median filter had the best performance against

TABLE 5.1
Mean Value of Comparison of PSNR, SNR, SSIM, and MSE, Parameters to 158 Images for Adaptive, Guided, Gaussian, Mean, and Median Filters in the Original Image Corrupted by Gaussian Noise (0.05)

Filter's name	MSE	PSNR	SNR	SSIM
Mean filter	0.00058	32.38	27.65	0.70
Median filter	0.00040	33.96	29.23	0.81
Adaptive filter	0.00026	36.16	31.42	0.90
Gaussian filter	0.0021	26.87	22.13	0.36
Guided filter	0.00073	31.36	26.62	0.62

TABLE 5.2
Comparison of PSNR, SNR, SSIM, and MSE Values for Adaptive, Guided, Gaussian, Mean, and Median Filters in the Original Image Corrupted by Salt-and-Pepper Noise (0.02)

Filter's name	MSE	PSNR	SNR	SSIM
Adaptive filter	0.00093	30.36	25.63	0.80
Guided filter	0.0024	26.24	21.51	0.61
Gaussian filter	0.0023	26.35	21.62	0.59
Mean filter	0.00066	31.84	27.11	0.75
Median filter	**4.96E-05**	**43.48**	**38.75**	**0.97**

TABLE 5.3
PSNR Evaluating Parameter Results for Different Filters Applied to Salt-and-Pepper Noise Images at Different Noise Levels

Salt-and-pepper noise %	10 Noise level	20 Noise level	30 Noise level	40 Noise level	50 Noise level
Adaptive filter	26.57	24.50	22.86	21.42	20.14
Guided Filter	17.70	13.87	11.73	10.26	9.150
Gaussian filter	19.32	16.25	14.43	13.12	12.10
Mean filter	24.70	21.40	19.37	17.85	16.64
Median filter	**39.51**	**34.16**	**31.08**	**28.52**	**24.55**

TABLE 5.4

PSNR Evaluating Parameter Results for Different Filters Applied to Gaussian Noise Images at Different Noise Levels

Gaussian noise %	10 Noise level	20 Noise level	30 Noise level	40 Noise level	50 Noise level
Adaptive filter	20.08	14.36	11.23	9.34	8.22
Guided filter	19.17	14.13	11.15	9.30	8.21
Gaussian filter	18.71	13.99	11.08	9.27	8.19
Mean filter	19.73	14.27	11.19	9.32	8.21
Median filter	**19.79**	**14.13**	**10.98**	**9.07**	**7.99**

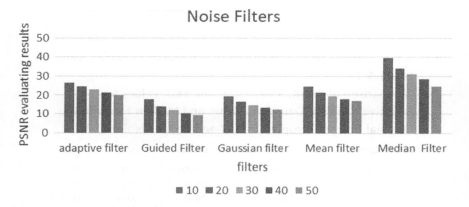

FIGURE 5.7   Bar chart showing the results of filtration applied to the original image at different levels of salt-and-pepper noise.

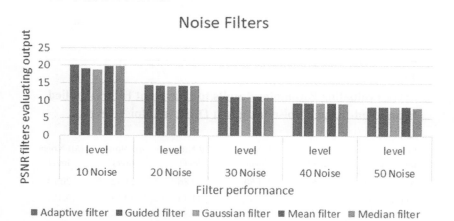

FIGURE 5.8   Bar chart showing the results of filtration applied to the original image at different levels of Gaussian noise.

salt-and-pepper noise at each noise level from 10% until 50%, as shown in Table 5.3. In Gaussian noise, the median filter and the adaptive filter performance were also close, as shown in Table 5.4. Guided by other studies and experiments (Geetha *et al.*, 2017; Priyadharshini *et al.*, 2014; Anindita Septiarini *et al.*, 2018; Sundari B. *et al.*, 2017; Claro M. *et al.*, 2016), we can conclude that the median filter performs better than other filters in removing noise and preserving boundaries in digital fundus images.

5.2.4 OPTIC DISC SEGMENTATION STEP

The optic disc is the pallor circular region located at the position where the optic nerve leaves the eye, where the optic cup is the central, bright, yellowish circular region in the optic disc. Optic disc segmentation: the proposed method can be divided into 3 steps. The **first step** is to remove the vessel to achieve accurate segmentation, and it is done by morphological operation. The **second step** is thresholding and binarization, and the given image will convert to a binary image. From this image, the optic disc boundary can be easily extracted.

Morphological operations are used to smooth the obtained disc and cup boundary, followed by labeling it as a binary image to get the disc and cup boundary. The **third step** is a disc reconstruction as a circle, using the center and radius from the detecting region, as discussed in Chapter 3.

This section describes the OD segmentation results. Figure 5.9 and Figure 5.10 show an example for the OD segmentation algorithm performance.

To choose the best OD segmentation methods, three segmentation methods have been tested. The results have been compared with each other, as shown in Figure 5.11.

The proposed algorithm was evaluated based on the following parameters:

The three parameters are: Dice similarity coefficient (DSC), Jaccard coefficient and Structural Similarity (SSIM)), where these parameters are common and accurate in evaluating the segmentation and used before by (F. Fumero *et al.*, 2015). Here it applied to compare the results of the segmentation algorithm with the five expert's see Section 4.2.1 describe the ground truth segmentation, the segmentation results show in Table 5.5, and then compared with other segmentation methods applied by other researcher to evaluate the segmentation algorithm performance, the results are shown in Table 5.6 and Figure 5.12.

Below are some examples that show the absolute difference between the same segmented images by the proposed method and the ground truth in Figures 5.13 and 5.14.

5.2.4.1 OD Discussion

An automatic algorithm for optic disc segmentation has been tested before (Fauzia *et al.*, 2013; Mohd *et al.*, 2017). Here, the researchers applied their algorithm to 158 healthy and glaucomatous images; 148 images were segmented correctly, the remaining 10 were not segmented duo to non-uniform illumination, the main blood vessel, and the large size of the PPA, which affected the disc boundary, as shown in Figure 5.15.

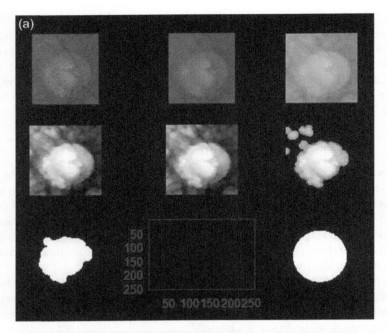

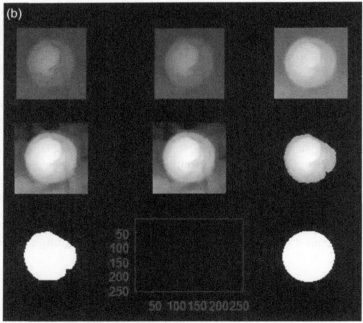

FIGURE 5.9 Shows a glaucomatous image's OD segmentation steps as vessel removal, thresholding, post-processing, and circle reconstruction. (a) Wrong OD segmentation due to PPA affect, and (b) correct OD segmentation.

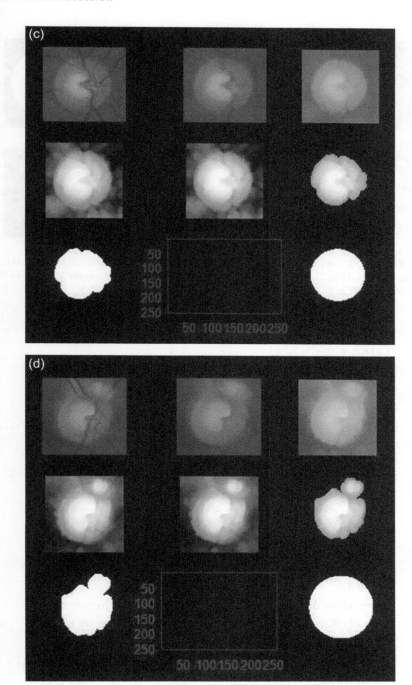

FIGURE 5.10 Shows a healthy image's OD segmentation steps as vessel remov"al, thresh-olding, post-processing, and circle reconstruction. (c) Correct OD segmentation, and (d) wrong OD segmentation.

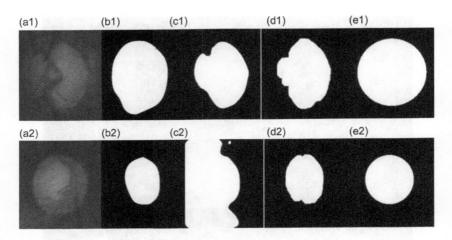

FIGURE 5.11 Shows comparison results of different segmentation algorithms. (a1, 2) Original image (b1, 2) ground truth image (c1, 2) Otsu Thresholding (d1, 2) Global Thresholding (e1, 2). Proposed method.

TABLE 5.5

The Mean Value of the DSC, Jaccard Coefficient, and SSIM Acquired Using 158 Images Segmented By the Proposed Method and Compared with the Five Ground Truths in the Database

Performance parameters %	Ground truth 1	Ground truth 2	Ground truth 3	Ground truth 4	Ground truth 5
DSC	84	88	**90**	87	89
Jaccard coefficient	73	79	**82**	82	81
SSIM	81	83	**83**	82	82

TABLE 5.6

Comparison of the Proposed Method with Other Segmentation Methods (Otsu Thresholding, Global Thresholding) by Using the Same Database and Evaluation Parameters

Segmentation method	DSC %	Jaccard coefficient %	SSIM %
Otsu thresholding	87	79	82
Global thresholding	85	75	82
Proposed method	**90**	**82**	**83**

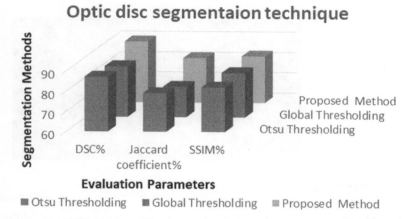

FIGURE 5.12 Bar chart showing a comparison between three different types of segmentation applied for disc segmentation to the same database.

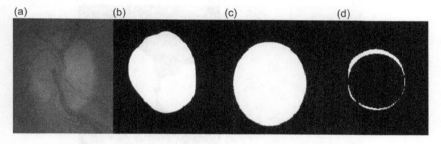

FIGURE 5.13 Shows accurate segmentation of: (a) the original image (b) the ground truth OD (c) proposed OD (d) the absolute difference between the two images is too small.

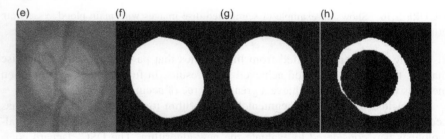

FIGURE 5.14 Shows inaccurate segmentation (e) the original image (f) ground truth OD (g) proposed OD (h) absolute difference between the two images is too big.

The reason of circle OD reconstruction step is to remove the un-accurate affect resultant from the main blood vessel, like in Figure 5.16, which shows the segmentations result before the circle OD reconstruction step.

The only limitation of this method is that each type of image needs a different thresholding level, in comparison with Sirshad *et al.*, (2016), which segmented the

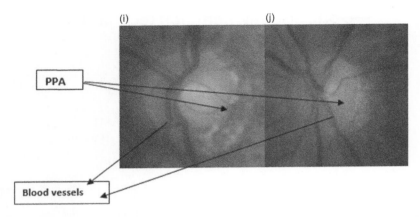

FIGURE 5.15 Shows an example of non-segmented images due to some reasons like (i) large size PPA (j) non-uniform illumination.

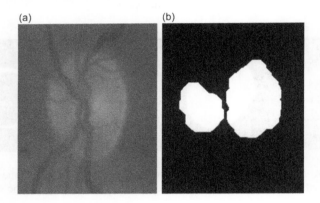

FIGURE 5.16 Shows an example of segmented disc images with main blood vessels (a): original image, (b) segmentation output before circle construction.

disc using a gradient extracted from line profiles that pass through the optic disc margin; the proposed method achieved good results. In future, other segmentation methods will be tested to achieve a greater degree of accuracy.

In conclusion, a new OD segmentation algorithm for digital fundus images has been developed based on the RIM_ONE dataset. The ground truth of five ophthalmologists was considered to evaluate the new algorithm. The DSC, Jaccard coefficient, and SSIM parameters were chosen to evaluate the new system in this paper and found that the best DSC was 90%, Jaccard 82%, and SSIM 83%.

5.2.5 OPTIC CUP SEGMENTATION STEP

Cup detection from digital fundus images is one of the most challenging step because it interferes with the blood vessels. Based on previous studies, the following segmentation methods are suggested: thresholding from the disc area, in order to obtain a greater degree of segmentation accuracy; the suggested optic cup segmentation method, depend on the optic disc thresholding at level 240, a clear border, and a smoothing boundary (and binaries) for the final image. The proposed algorithm method is shown in Chapter 4.

In this section, a cup boundary detection algorithm is presented and the segmentation results from the digital fundus images are evaluated in the DRISHTI-GS database. An example of the segmentation is shown in Figure 5.17.

5.2.5.1 Experimental Results

The proposed cup segmentation method was evaluated by three parameters, which are: dice coefficient, Jaccard coefficient, and SSIM mentioned in Section 3.3. The results were compared with other methods like super pixel segmentation method, modified super pixel, modified thresholding, and the best results obtained by the global thresholding technique as shown in Figures 5.18 and 5.19.

The cup segmentation using the global thresholding technique was applied and compared with other segmentation techniques. The results were evaluated based on standard parameters and can be seen in Figures 5.19 through 5.21 and also in Table 5.7.

5.2.5.2 OC Discussion

Although experimentally, it was demonstrated that the use of global thresholding can help improve optic cup detection; however, it should be noted that there are some limitations and considerations for the effective use of global thresholding for cup boundary detection, like the thresholding level differentiating from one data base to another. Even for images in one database, there were several instances of detection

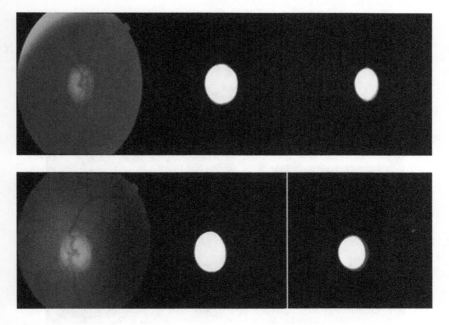

FIGURE 5.17 Sample images and ground truth in DRISHTI_GS dataset (from left to right). Original image (2,047 × 1,751) in the dataset, ground truth optic disc mask, proposed segmented optic cup mask.

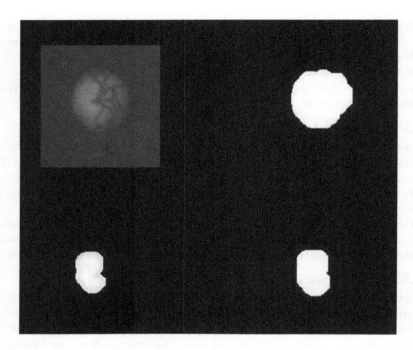

FIGURE 5.18   Shows cup segmentation and binarization from the disc part of a glaucomatous image using the global thresholding technique.

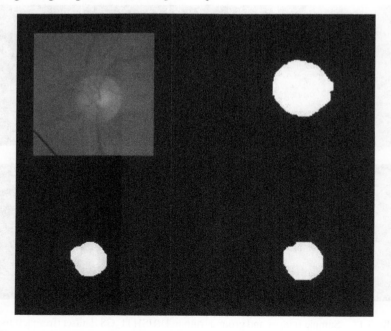

FIGURE 5.19   Shows cup segmentation and binarization from the disc part of a healthy image using the global thresholding technique.

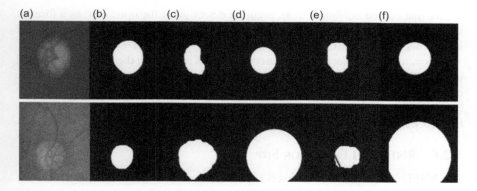

FIGURE 5.20 Shows (a) the original image (b) ground truth image (c) super pixel segmentation (d) super pixel segmentation with circle construction (e) global thresholding (f) global thresholding segmentation with circle construction.

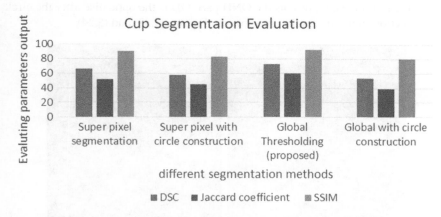

FIGURE 5.21 Bar chart illustrating the comparison between different segmentation techniques.

TABLE 5.7

Comparison of the Proposed Method with Other Segmentation Methods (Super Pixel Segmentation, Global Thresholding, Super Pixel with Circle Construction, and Global with Circle Construction) by Using the Same Database and Evaluation Parameters

Segmentation method	DSC %	Jaccard coefficient %	SSIM %
Super pixel segmentation	66	52	90
Super pixel with circle construction	58	45	83
Global thresholding (proposed)	**73**	**60**	**93**
Global with circle construction	53	39	80

failure that can be attributed to an accurate detection of the optic disc as a form of initialization for the optic cup detection. Figure 5.22 is an example of un-segmented images due to bad localization of the ROI region.

In conclusion, this section has presented models to localize the optic cup in retinal images. The framework provides two major contributions. Firstly, robust and stable optic cup localization was presented. Secondly, an accurate, simple, and inexpensive computation method was proposed. Experimentally, it was demonstrated that this method was able to produce optic cup localization with high accuracy.

5.2.6   RNFL ROI Extraction Step

The RNFL (retinal nerve fiber layer) is the area surrounding the disc for the ROI. It is based on a mathematical method to subtract the whole image from the disc area, by the formula:

$$RNFL = ROI\ image - OD \qquad (5.1)$$

Where the ROI image represents the ONH part, OD is the optic disc after the circle reconstruction step. The results are shown in Figures (5.23) and (5.24).

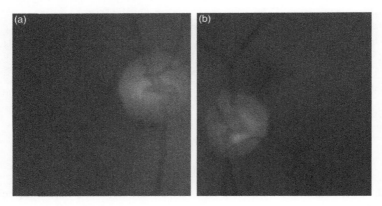

FIGURE 5.22   An example of an unsegmented image due to bad ROI localization.

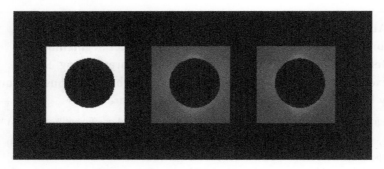

FIGURE 5.23   Shows the RNFL for a glaucomatous image, which is in part surrounded by the disc.

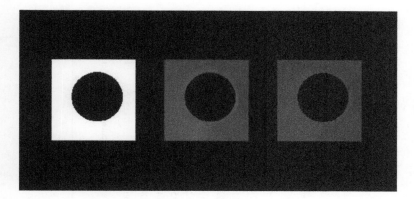

FIGURE 5.24 Shows the RNFL for a healthy image, extracted by the optic nerve head from the segmented optic disc.

The OD, OC, and RNFL are used for the third step, which is feature extraction and selection, and the final step is the classification.

5.3 SECTION TWO: GLAUCOMA DETECTION

5.3.1 FEATURE EXTRACTION STEP

Depending on an ophthalmologist's observation, glaucoma is diagnosed by an examination of the size, structure, shape, and color of the optic nerve head and RNFL. Based on this theory, a combination of size, structure, and shape features are applied to the disc, cup, and RNFL to classify glaucomatous from non-glaucomatous cases. These features are:

5.3.1.1 Shape Features

Measurements of several properties describe the shape for each disc and cup within the region props function. This computes area, convex area, eccentricity, centroid, filled area, filled image, extent, major axis length, minor axis length, solidity, radius, and perimeter. These measurements are applied for disc and cup masks, and the results are shown in Tables 5.8 and 5.9, which have been calculated by MATLAB® (2016).

All these shape features applied to the t-test to obtain the most significant features are based on: if P-value $< \alpha = 0.10$, and accepting the null hypothesis at $\alpha > 0.10$, means that there is no significant difference between variables. The final results from Table 5.10 show that there is a significance of 90% in the cup minor axes and disc solidity features calculated using SPSS software.

From the table above, one significant feature is the cup minor axes. The results are shown in Figure 5.25 and Table 5.11.

5.3.1.2 Color Features

Color moments are measurements that characterize color distribution in an image in the same way that central moments uniquely describe a probability distribution.

TABLE 5.8

Illustrates Examples Cup Shape Features for Only 10 Images

CON	ECC	EQU	EX	F.REA	MAJ	MIN	OR	PER	SOL	C.AREA	C.C.X	C.C.Y
7734	0.51365	98.2921	0.79017	7588	106.713	91.5592	−36.166	314.046	0.98112	7588	139.057	125.206
8545	0.61837	103.338	0.84717	8387	117.596	92.417	79.3882	337.334	0.98151	8387	112.914	122.86
8979	0.65156	104.386	0.7703	8558	121.962	92.5199	−59.811	348.068	0.95311	8558	149.857	110.953
8352	0.54682	102.017	0.80931	8174	121.217	93.9534	47.6436	328.262	0.97869	8174	120.66	130.916
8462	0.52967	102.905	0.79971	8317	112.108	95.0897	−51.553	329.172	0.98286	8317	140.109	122.702
8602	0.65355	102.887	0.75438	8314	119.375	90.353	−42.763	341.826	0.96652	8314	142.535	116.383
8559	0.55092	103.682	0.80517	8443	113.797	94.9702	59.6989	329.85	0.98645	8443	98.7616	117.397
8079	0.6285	100.717	0.86786	7967	114.812	89.3026	−85.403	323.48	0.98614	7967	98.9665	127.523
8594	0.48608	103.59	0.83545	8428	112.069	97.9385	27.3424	336.904	0.98068	8423	110.236	120.323
8605	0.51489	103.866	0.80665	8473	112.913	96.7953	52.5652	332.24	0.98466	8473	115.999	100.395

Where cup features are: CON = convex area, EQU = equivalent diameter, EX = extent, MAJ = major axes, MIN = minor axes, OR = orientation, PER = perimeter, SOL = solidity, F.AREA = filled area, ECC = Eccenterity, C.AREA = cup area, C.C.Y = cup center y, C.C.X = cup center x.

TABLE 5.9
Illustrates Examples of Disc Shape Features for Only 10 Images

D.AREA	D.C.X	D.C.Y	D.rad.	CON	ECC	EQU	EX	F.ARE	MAJ	OR	MIN	PER	SOL	D.RAD
18582	132.25	128.139	96.1361	18705	0.04091	153.816	0.78352	18582	153.885	−20.743	153.756	480.392	0.993422	56.0946
16430	126.982	127.648	90.4076	16546	0.03836	144.635	0.78145	16430	144.587	−15.293	144.587	451.918	0.99299	64.1524
19755	129.304	118.283	94.5063	18069	0.02812	151.199	0.78747	17955	151.173	−46.992	151.173	471.94	0.99369	78.6076
17700	122.841	126.364	93.8218	17809	0.02418	150.121	0.78667	17700	150.103	−85.056	150.103	468.392	0.99388	59.7453
16800	123.275	115.251	91.414	16901	0.02691	146.255	0.78814	16800	146.233	51.2652	146.233	456.21	0.99402	60.8713
19378	135.66	115.769	98.1773	19497	0.024	157.076	0.78616	19378	157.058	−53.318	157.058	490.49	0.9939	65.1853
18543	113.516	120.333	96.014	18663	0.02615	153.654	0.78188	18543	153.632	8.29331	153.632	480.02	0.99357	63.8804
18901	112.982	127.223	96.927	19015	0.04101	155.131	0.78672	18901	155.069	−5.2522	155.069	484.122	0.994	71.229
17614	124.792	124.994	93.6193	17722	0.0375	149.756	0.7881	17614	149.708	7.91116	149.708	467.54	0.99391	59.0958

Where disc feature are: CON = convex area, EQU = equivalent diameter, EX = extent, MAJ = major axes, MIN = minor axes, OR = orientation, PER = perimeter, SOL = solidity, F.AREA = filled area, ECC = Eccenterity, C.C.Y = cup center y, C.C.X = cup center x, D.AREA = disc area, D.RAD = disc radius.

TABLE 5.10
Shows the t-test Results from the Shape Features

Variables	Pictures	N	Mean	Std. Dev.	S.E.M.	F	t.	df.	P. Value $\alpha > 0.10$
C. CON	Infect	36	8.730	1.01	1.07	0.029	0.841	151	0.40
	Control	117	8. 570	9.95	9.16				
C.ECC	Infect	36	0.550	0.132	0.02	0.108	1.209	151	0.23
	Control	117	0.580	0.134	0.01				
C.EQU	Infect	36	1.040	4.17	0.71	0.25	0.934	151	0.35
	Control	117	1.030	5.58	0.51				
C.EXTENT	Infect	36	0.880	0.044	0.01	0.526	0.718	151	0.47
	Control	117	0.8130	0.039	0				
C.F.AREA	Infect	36	8.480	7.18	1.22	0.113	0.909	151	0.37
	Control	117	8.340	8.48	7.81				
C.MAJ. AXES	Infect	36	1.160	1.15	1.95	0.005	0.245	151	0.81
	Control	117	1.170	1.25	1.15				
C.MIN. AXES	Infect	36	9. 490	4.96	0.839	2.24	1.937	151	0.03
	Control	117	9.250	6.99	0.643				
C.OREN	Infect	36	1.630	5.93	1	4.695	0.739	151	0.46
	Control	117	7.030	6.72	6.18				
C.PER	Infect	36	3.410	4.15	7.02	0.868	0.777	151	0.44
	Control	117	3.360	3.02	2.78				
C.SOL	Infect	36	0.974	0.022	0.004	0.935	0.11	151	0.91
	Control	117	0.975	0.025	0.002				
C. AREA	Infect	36	8.480	7.2	1.25	0113	0.909	151	0.37
	Control	117	8.340	8.48	7.81				
C.CENT ERX	Infect	36	1.230	1.99	3.38	0.004	0.96	151	0.34
	Control	117	1.270	2.11	1.95				
C.CENTRY	Infect	36	1.230	1.16	1.95	1.667	0.031	151	0.98
	Control	117	1.230	1.1	1.01				
D.AREA	Infect	36	1.870	1.31	2.22	5.881	0.932	151	0.35
	Control	117	1.830	2.68	2.47				
D.CENT ERX	Infect	36	1.230	1.01	1.71	0.636	−0.394	151	0.17
	Control	117	1.260	9.62	0.886				
D.CENTRY	Infect	36	1.250	8.06	1.36	0.002	0.611	151	0.54
	Control	117	1.240	8.49	0.78				
D. Red	Infect	36	9.650	3.36	0.57	5.891	1.065	151	0.29
	Control	117	9.510	3.33	0.67				
D.CON	Infect	36	1.880	1.32	2.22	5.891	0.478	151	0.63
	Control	117	1.840	2.69	2.48				
D.ECC	Infect	36	0.034	0.009	0.002	0.018	1.063	151	0.29
	Control	117	0.033	0.01	0.001				
D.EQU	Infect	36	1.540	5.36	0.907	5.809	0.929	151	0.35
	Control	117	1.520	1.17	1.08				

(Continued)

TABLE 5.10 (CONTINUED)
Shows the t-test Results from the Shape Features

Variables	Pictures	N	Mean	Std. Dev.	S.E.M.	F	t.	df.	P. Value $\alpha > 0.10$
D.EX	Infect	36	0.785	0.002	0	0.359	0.333	151	0.74
	Control	117	0.785	0.003	0				
D.F.AREA	Infect	36	1.870	1.31	2.22	5.881	0.932	151	0.35
	Control	117	1.830	2.68	2.47				
D.MAJ. AXES	Infect	36	1.540	5.37	0.91	5.893	1.063	151	0.29
	Control	117	1.520	1.17	1.08				
D.MIN. AX	Infect	36	1.540	5.36	0.906	5.924	1.062	151	0.29
	Control	117	1.520	1.17	1.08				
D.OR	Infect	36	8.810	4.86	8.22	0.391	0.265	151	0.79
	Control	117	6.170	5.26	4.84				
D.PER	Infect	36	4.820	1.68	2.83	0.94	1.057	151	0.29
	Control	117	4.750	3.68	3.39				
D.SOL	Infect	36	0.994	0	0	0.667	1.881	151	0.05
	Control	117	0.993	0	0				
D.RAD	Infect	36	6.590	7.46	1.26	0.069	0.058	151	0.95
	Control	117	6.580	1.26	1.157				

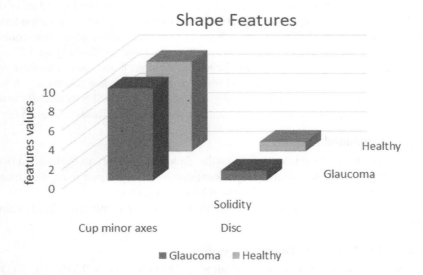

FIGURE 5.25 Bar chart showing the cup minor axes and disc solidity for 40 glaucomatous images and 118 healthy images.

TABLE 5.11

The Selected Shape Features for the

Glaucomatous and Healthy Images

Digital fundus images	Cup minor axes	Disc solidity
Glaucomatous	9.500	0.994
Healthy	9.300	0.993

Color moments are mainly used for color indexing purposes as features in image retrieval applications in order to compare how two similar images are based on color. Three color moments are computed per channel, nine color features are computed for each disc, cup, and RNFL, and these features are mean, standard deviation, and skewness (discussed in Chapter 2). The results are shown in Tables 5.12 through 5.14.

SPSS software was applied to the color features in the above tables to choose the most significant features using the t-test method. The results are shown in Tables 5.15 and 5.16, and illustrate that there are five significant color features for the cup mean, cup STD, disc mean, disc STD, and RNFL. STD is shown in Figure 5.26.

5.3.1.3 Texture Features

GLCM and Tamura algorithms can be used to describe the main features of the textures, such as coarseness and regularity. The gray-level co-occurrence matrices are one of the most important measurements that can be used to describe the texture of the RNFL. The features obtained are 22 features from the GLCM at 32 quantization levels based on Sakthivel et al., (2014), and 3 features from the Tamura algorithm (coarseness, direction, and contrast). The results are shown in Table 5.17 and Figure 5.27.

After the application of the t-test for the texture features, it was found that there was no significant difference. After that, and based on previous studies (Chapter 3), it was proven that the textures are affected by the glaucoma disease. The SFS method discussed in Chapter 2 was applied and coarseness features were selected (Tables 5.18 and 5.19).

5.3.1.4 Combined Features

From the above section, it is apparent that the glaucoma disease can be detected from fundus images via eight features (cup mean, cup STD, cup minor axes, disc mean, disc STD, disc solidity, RNFL STD, and RNFL coarseness).

$R = 0.245$, $R^2 = 60\%$, where R represent the effect of the features on the glaucoma by 60%.

5.3.1.5 The Glaucoma Linear Regression

$= 419.460 + OD$ mean. $\times 3.962 + OC$ mean $\times -2.329 + OD$ STD $\times 0.340 + OC$ STD $\times -2.794 + OC$ minor axis $\times 0.309 + RNFL$ coarseness $\times -1.201$

Where mean, standard deviation, minor axes, and coarseness are explained in Sections 2.8.6, 2.9.1, 2.9.2, and 2.11.1.

TABLE 5.12
Illustrates Examples of Cup Color Features for Only 10 Images

c.meanb	c.meang	c.meanr	c.skb	c.skg	c.skr	c.stdb	c.stdg	c.stdr
7.90181	14.0489	25.2425	2.54386	2.58408	2.40448	22.1806	39.739	69.7791
2.09569	6.54398	20.0834	3.41193	3.21695	2.30878	6.22521	19.4927	52.9064
3.24767	9.00752	24.5948	2.95347	2.60921	2.24419	9.11699	25.8478	63.8321
6.93747	12.5416	26.2109	2.51523	2.57970	2.28106	18.8929	34.5650	69.5096
7.72103	14.2797	27.5546	2.40265	2.49326	2.24777	2.61168	38.6886	72.3233
5.84283	11.7086	28.7879	2.40234	2.59194	2.24719	15.6165	32.3337	75.5673
7.67693	14.3331	28.8349	2.31481	2.43283	2.22345	20.2012	38.4693	75.0494
9.21680	17.2846	29.3722	2.45051	2.486167	2.320597	25.1438	47.55311	78.99562
1.488968	6.372147	21.9944	2.648844	2.648259	2.252229	4.063342	17.53406	57.46885

Where c.meanb = cup mean for blue channel, c.meanr = cup mean for red channel, c.meang = cup mean for green channel, c.skg = cup skewness for green channel, c.skb = cup skewness for blue channel, c.skr = cup skewness for red channel, c.stdb = cup standard deviation for blue channel, c.stdr = cup standard deviation for red channel, c.stdg = cup standard deviation for green channel.

TABLE 5.13
Illustrates Examples of Disc Color Features for Only 10 Images

d.meanb	d.meang	d.meanr	d.skb	d.skg	d.skr	d.stdb	d.stdg	d.stdr
17.1859	29.90857	5943636	1.124948	1.20395	0.970426	27.99239	49.53667	94.6255
3.295288	10.19006	36.144	2.453966	2.314909	1.25002	6.786466	21.2088	63.3845
5.321747	14.44046	47.60818	1.98130	2.51152	1.08890	9.934429	28.0028	78.3869
14.0538	22.9087	54.2377	1.32454	1.45607	1.05368	22.4091	40.5190	89.4100
14.0824	25.1448	53.6949	1.29562	1.43360	1.12720	24.5997	45.0469	91.6031
12.4003	24.2752	64.2054	1.05455	1.25686	0.91911	19.6200	40.1474	99.2040
15.5842	28.0077	61.2039	1.10117	1.26401	0.97591	25.3640	47.1826	97.6161
19.2535	34.6776	67.4584	1.10083	1.23425	0.94176	30.9907	57.5081	106.085
2.51455	10.7772	42.2685	1.64029	1.66259	1.10318	4.52128	19.7808	70.3417
13.4287	3008844	61.6237	1.30698	1.27352	0.91480	22.0351	49.8715	95.0597

Where d.meanb = disc mean for blue channel, d.meanr = disc mean for red channel, d.meang = disc mean for green channel, d.skg = disc skewness for green channel, d.skb = disc skewness for blue channel, d.skr = disc skewness for red channel, d.stdb = disc standard deviation for blue channel, d.stdr = disc standard deviation for red channel, d.stdg = disc standard deviation for green channel.

5.4   SECTION THREE: CLASSIFICATION

The classification performance assessment was carried out with the Receive Operating Characteristics (ROC) Curve. A binary classification model classifies each instance into one of two classes; say *glaucomatous* and a *non-glaucomatous* class. This gives rise to four possible classifications for each instance: a true positive (TP), a true

TABLE 5.14

Illustrates Examples of RNFL Color Features for Only 10 Images

rn.meanb	rn.meang	rn.meanr	rn.skb	rn.skg	rn.skr	rn.stdb	rn.stdg	rn.stdr
33.04367	57.96133	120.3777	−0.83114	−0.7216	−0.8025	21.19177	37.83722	77.59473
5.337463	21.27376	65.88588	0.850968	−0.2451	−0.3573	4.217105	14.15665	43.54503
6.793976	24.85201	74.96664	−0.07356	−0.2465	−0.2820	4.911454	17.02542	51.91936
29.0159	48.4364	111.5854	−0.89974	−0.6311	−07241	18.00881	31.37594	71.37176
32.28215	55.11308	116.7324	−086873	−05037	−0.7154	19.63159	35.34451	73.0528
25.38705	47.01894	117.8963	−075201	−0.5583	−0.6225	16.82775	32.11181	79.7713
32.25183	57.31044	112.6929	−0.83092	−0.6127	−0.5560	20.68434	38.04992	75.70935
39.45503	67.35126	129.4043	−0.8721	−0.6692	−0.6662	25.34794	44.59389	85.97335
3.712463	23.88306	84.02716	−0.62445	−0.8250	−0.8481	2.41081	14.94325	52.42571
19.39464	37.93088	104.9089	−0.75676	−0.4434	−0.5161	12.88353	26.37245	72.68975

Where rn.meanb = RNFL mean for blue channel, rn.meanr = RNFL mean for red channel, rn.meang = RNFL mean for green channel, rn.skg = RNFL skewness for green channel, rn.skb = RNFL skewness for blue channel, rn.skr = RNFL skewness for red channel, rn.stdb = RNFL standard deviation for blue channel, rn.stdr = RNFL standard deviation for red channel, rn.stdg = RNFL standard deviation for green channel.

negative (TN), a false positive (FP), or a false negative (FN). This situation can be depicted as a confusion matrix (also called a contingency table) given in Figure 5.28. The confusion matrix illustrates the observed classifications for a phenomenon (columns) with the predicted classifications of a model (rows). In Figure 5.35, the classifications that lie along the major diagonal of the table are the correct classifications; that is, the true positives and the true negatives. The other fields signify model errors. For a perfect model the true positive and true negative fields will be filled out, and the other fields will be zero. The performance metrics can be calculated from the confusion matrix fields.

ROC curves were originally designed as tools to visually determine optimal operating points for the classifier. Two new performance metrics need to be introduced to construct ROC curves (defined here in terms of the confusion matrix), the true positive rate (TPR) and the false positive rate (FPR).

ROC graphs are constructed by plotting the true positive rate (TPR) against the false positive rate. Figure 5.29 identifies a number of regions of interest in an ROC graph. The diagonal line from the bottom-left corner to the top-right corner shows the classifier's performance. In the extreme case, denoted by the point in the bottom-left corner, a conservative classification model will classify all instances as negative and it will not commit any false positives, but it will also not obtain any true positives. The region of classifiers' performance appears at the top of the graph. These classifiers have a good true positive rate, but they also have a number of false positive errors. When the classifier is at the top-right corner it means that it classifies every instance as positive. In this situation, the classifier will not miss any true positives, but it will also miss a very large number of false positives. If the classifiers fall to the right of the random performance line, this mean it has a performance worse

TABLE 5.15
Shows the t-test Results from the Color Features

Variables	Pictures	N	Mean	Std. Dev.	S.E.M.	F	t.	df	P.Value $\alpha > 0.10$
C. meanb	Infect	35	6.22	2.6	0.44	0.225	0.786	151	0.43
	Control	118	6.76	3.76	0.35				
C. meang	Infect	35	12.19	3.48	0.59	0.332	0.121	151	0.90
	Control	118	12.11	3.67	0.34				
c.meanr	Infect	35	26.16	3.31	0.56	0.22	2.588	151	0.01
	Control	118	24.35	3.72	0.34				
c.stdb	Infect	35	18.66	6.86	1.16	0.041	0.853	151	0.40
	Control	118	18.2	9.95	0.92				
c.stdg	Infect	35	32.15	8.99	1.52	0.528	0.099	151	0.92
	Control	118	32.97	9.31	0.86				
c.stdr	Infect	35	68.07	8.22	1.39	4.303	2.802	151	0.01
	Control	118	63.77	7.9	0.73				
c.skb	Infect	35	2.54	0.32	0.05		0.516	151	0.61
	Control	118	2.57	0.37	0.03				
c.skg	Infect	35	2.59	0.27	0.05		0.08	151	0.94
	Control	118	2.58	0.27	0.02				
c.skr	Infect	35	2.23	0.13	0.02		0.998	151	0.32
	Control	118	2.27	0.22	0.02				
d. meanb	Infect	35	12.79	7.64	0.7	0.1	0.644	151	0.64
	Control	118	23.4	8.02	1.35				
d.meang	Infect	35	23.4	8.02	1.35	0.198	0.277	151	0.78
	Control	118	22.96	8.37	0.77				
d.meanr	Infect	35	55.1	9.15	1.55	3.293	2.33	151	0.02
	Control	118	50.61	10.53	0.94				
d.skb	Infect	35	1.35	0.4	0.07	0.3	0.914	151	0.36
	Control	118	1.43	0.46	0.04				
d.skg	Infect	35	1.44	0.35	0.06	0.072	0.157	151	0.88
	Control	118	1.45	0.36	0.03				
d.skr	Infect	35	0.99	0.12	0.02	0.305	1.261	151	0.21
	Control	118	1.04	0.26	0.02				
d.stdb	Infect	35	20.2	8.67	1.47	0.021	0.646	151	0.52
	Control	118	21.6	11.94	1.1				
d.stdg	Infect	35	39.53	11.85	2	4.23	0.25	151	0.80
	Control	118	38.97	11.75	1.08				
d.stdr	Infect	35	87.29	12.4	2.1	3.002	2.81	151	0.01
	Control	118	81.13	11.08	1.02				
RN.meanb	Infect	35	21.64	11.34	1.92	0.026	0.264	151	0.79
	Control	118	20.99	13.15	1.21				
RN.meang	Infect	35	41.75	14.86	2.51	1.417	0.523	151	0.60
	Control	118	40.26	14.74	1.36				
RN.meanr	Infect	35	101.61	20.25	3.42	0.659	1.226	151	0.22
	Control	118	97.33	17.47	1.62				

(Continued)

TABLE 5.15 (CONTINUED)
Shows the t-test Results from the Color Features

Variables	Pictures	N	Mean	Std. Dev.	S.E.M.	F	t.	df	P.Value $\alpha > 0.10$
RN.skb	Infect	35	0.49	0.45	0.08	1.96	1.564	151	0.12
	Control	118	0.05	1.65	0.15				
RN.skg	Infect	35	0.48	0.25	0.04	3.151	0.856	151	0.39
	Control	118	0.42	0.41	0.04				
RN.skr	Infect	35	0.64	0.18	0.03	3.632	1.392	151	0.16
	Control	118	0.7	0.25	0.02				
RN.stdb	Infect	35	14.43	7.29	1.23	6.844	0.113	151	0.91
	Control	118	14.25	8.46	0.78				
RN.stdr	Infect	35	76.48	12.97	2.19	0.755	2.019	151	0.07
	Control	118	63.12	10.65	0.98				

TABLE 5.16
Illustrates the Mean Values of the Selected Color Features for Both 40 Glaucomatous and 118 Healthy Images

Digital fundus image	Cup (mean)	Cup (STD)	Disc (mean)	Disc (STD)	RNFL (STD)
Glaucomatous	26.04	67.71	54.78	86.85	67.08
Healthy	24.38	63.84	50.66	81.20	63.20

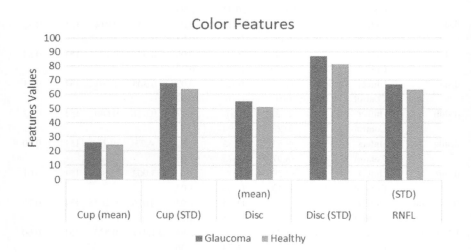

FIGURE 5.26 Bar chart of selected color features for 40 glaucomatous images and 118 healthy images.

TABLE 5.17
Illustrates Examples of RNFL Texture Features for Only 10 Images

COAR.	Cont.	DIR.	V1	V10	V11	V12	V13	V14
38.41763	40.69872	0.782514	8	0.828125	0.8075	0.714286	14.42212	4.626
38.46411	17.86302	0.062312	8.125	0.953125	0.9475	0.839286	11.05884	3.875
37.43988	21.51254	0.788693	8.125	0.953125	0.9475	0.839286	11.05884	3.875
38.7186	35.12054	0.788262	7.875	0.84375	0.825	0.732143	13.30103	4.5
38.84473	37.03552	0.79052	7.875	0.84375	0.825	0.732143	13.30103	4.5
38.18777	38.56587	0.807517	10.375	0.8125	0.79	0.660714	16.66431	5.25
38.41445	40.5225	0.780941	8	0.828125	0.8075	0.714286	14.42212	4.626
38.37451	46.92262	0.785659	9.375	0.796875	0.7725	0.660714	16.66431	5.125
38.84955	20.58708	4.59394	7.125	0.9375	0.93	0.839286	9.937744	3.75
38.29657	33.39359	0.761134	8.875	0.859375	0.8425	0.732143	14.42212	4.625

Where, COAR. = coarseness, Cont. = contrast, Dir. = direction, V1 = Energy
V10 = Cluster Prominence, V11 = Maximum probability, V12 = Sum of Squares, V13 = Sum Average,
V14 = Sum Variance.

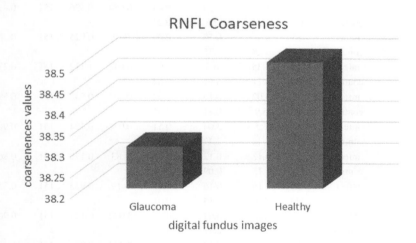

FIGURE 5.27 Shows the different means between glaucomatous and healthy RNFL coarseness features.

than random performance, due to it producing more false positive than true positive responses. However, because ROC graphs are symmetrical along the random performance line, the point in the top-left corner denotes perfect classification: 100% true positive rate and 0% false positive rate.

Figure 5.30 shows some typical examples of ROC curves. Part (a) depicts the ROC curve of an almost perfect classifier where the performance curve almost touches the "perfect performance" point in the top-left corner. Part (b) and part (c) depict ROC curves of inferior classifiers.

TABLE 5.18
Shows the t-test Results from the Texture Features

Variables	Pictures	N	Mean	Std. Dev.	S.E.M.	F	t.	df	P.Value $\alpha > 0.10$
Coar	infect	36	3.84	0.42	0.07	8.815	0.951	151	0.34
	control	117	3.85	0.85	0.08				
Cont.	infect	36	3.28	9.04	1.53	1.289	1.282	151	0.20
	control	117	3.07	8.04	0.74				
Dir.	infect	36	3.47	1.27	0.21	0.005	0.299	151	0.77
	control	117	2.64	1.49	0.14				
v1	infect	36	7.95	1.57	0.27	2.646	1.171	151	0.24
	control	117	7.54	1.88	0.17				
v10	infect	36	8.66	0.05	0.01	3.032	0.124	151	0.90
	control	117	8.67	0.04	0				
v11	infect	36	8.5	0.06	0.01	3.032	0.124	151	0.90
	control	117	8.51	0.05	0				
v12	infect	36	7.54	0.07	0.01	0.472	0.65	151	0.52
	control	117	7.61	0.06	0.01				
v13	infect	36	1.32	2.59	0.44	0.084	0.826	151	0.41
	control	117	1.28	2.6	0.24				
v14	infect	36	4.38	0.56	0.09	0.01	0.925	151	0.36
	control	117	4.28	0.55	0.05				
v15	infect	36	3.35	6.3	1.06	0.648	1.112	151	0.27
	control	117	3.21	6.97	0.64				
v16	infect	36	7.02	0.13	0.02	0.863	0.542	151	0.59
	control	117	6.9	0.11	0.01				
v17	infect	36	7.5	3.04	0.51	0.032	0.124	151	0.90
	control	117	7.44	2.47	0.23				
v18	infect	36	4.13	0.12	0.02	4.83	0.176	151	0.86
	control	117	4.15	0.09	0.01				
v19	infect	36	2.16	0.16	0.03	3.19	0.935	151	0.35
	control	117	1.9	0.14	0.01				
v2	infect	36	7.5	3.04	0.51	3.032	0.124	151	0.90
	control	117	7.44	2.47	0.23				
v20	infect	36	3.1	0.12	0.02	0.433	0.875	151	0.38
	control	117	3.8	0.11	0.01				
v21	infect	36	9.29	0.03	0	3.032	0.124	151	0.90
	control	117	9.29	0.02	0				
v22	infect	36	9.34	0.03	0	3.032	0.124	151	0.90
	control	117	9.34	0.02	0				
v3	infect	36	4.79	0.18	0.03	1.505	0.866	151	0.39
	control	117	4.52	0.16	0.01				
v4	infect	36	4.79	0.18	0.03	1.505	0.866	151	0.39
	control	117	4.52	0.16	0.01				

(Continued)

TABLE 5.18 (CONTINUED)
Shows the t-test Results from the Texture Features

Variables	Pictures	N	Mean	Std. Dev.	S.E.M.	F	t.	df	P.Value $\alpha > 0.10$
v5	infect	36	1.77	370.37	62.6	0	1.101	151	0.27
	control	117	1.69	367.51	33.83				
v6	infect	36	1.47	28.95	4.89	0.065	1.016	151	0.31
	control	117	1.41	27.82	2.56				
v7	infect	36	1.07	0.43	0.07	3.032	0.124	151	0.90
	control	117	1.06	0.35	0.03				
v8	infect	36	5.96	0.09	0.02	0.994	0.567	151	0.57
	control	117	6.05	0.08	0.01				
v9	infect	36	7.95	0.18	0.03	1.907	0.392	151	0.70
	control	117	0.78	0.14	0.01				

TABLE 5.19
Shows the Selected Feature by SFS Method

Digital fundus image	RNFL coarseness
Glaucomatous	38.3
Healthy	38.5

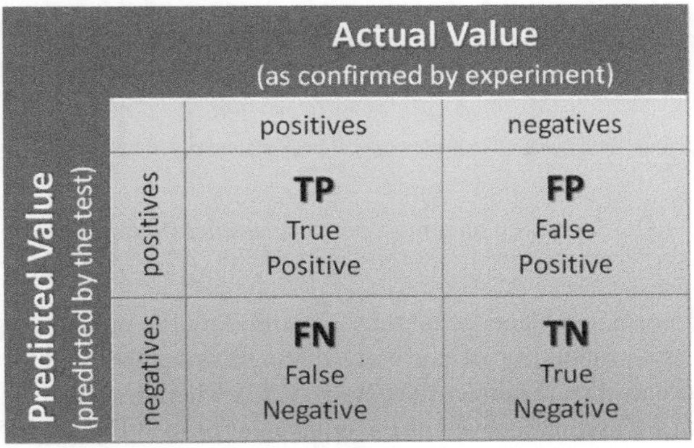

FIGURE 5.28   Format of a confusion matrix.

Here, the ROC curves used to check the performance of classifiers to distinguish between normal and glaucomatous fundus images, using the ROC curve parameters like sensitivity, specificity, and accuracy, where sensitivity indicates the number of subjects who have glaucoma and are accurately identified by the positive test. Thus, it is a measure of the probability of correctly diagnosed images. Specificity indicates

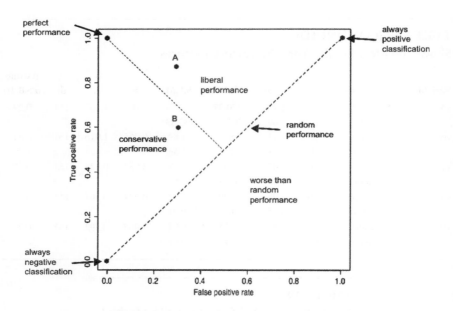

FIGURE 5.29   ROC curve: regions of a ROC graph.

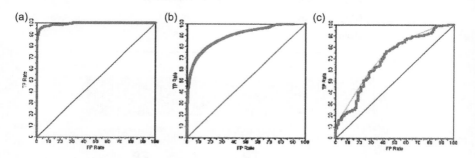

FIGURE 5.30   ROC curves: (a) an almost perfect classifier (b) a reasonable classifier (c) a poor classifier.

the number of normal subjects accurately identified by the negative test. Thus, it is a measure of the probability of correctly distinguishing the healthy images. The accuracy represents the area under the curve (AUC), where it represents the expected performance of a classifier, as well as the equivalent probability to randomly chosen positive instance higher than a randomly chosen negative instance.

- **True positive (TP)** is the number images detected as glaucomatous by an expert and the proposed method.
- **True negative (TN)** is the number of images detected as non-glaucomatous by an expert and the proposed method.
- **False positive (FP)** is the number of images detected as non-glaucomatous by an expert, but detected as glaucomatous by the proposed method.

- **False negative (FN)** is the number of images detected as glaucomatous by an expert, but detected as non-glaucomatous by the proposed method.

5.4.1 SHAPE FEATURES CLASSIFICATION

The selected features are cup minor axes and disc solidity. These features are evaluated by many classifiers like SVM, KNN, ensembles bagging classifier, and ensembles boosting classifier discussed in Section 2.13. The classification results in Figure 5.31 show the best result using SVM, which have an accuracy of 76.5% and AUC 0.052.

From the above results, TP, TN, FP and FN = 117, 0, 0, and 36, respectively, the researchers noticed that the healthy images are correctly classified (117) and all glaucomatous image are misclassified (36 of 36) due to imbalanced features. (When examples of one class in a training data set vastly outnumber examples of the other class, traditional data mining algorithms tend to create suboptimal classification models.) To solve this problem, the smote a logarithm discussed in Section 2.14 and used in (kumar *et al.*, 2012; Alghamdi *et al.*, 2017; Yuan *et al.*, 2015) was applied to balance the features; here the balanced features are classified again and the results are shown in Figure 5.32 to obtain the following: TP, TN, FP and FN = 40, 134, 15, and 7, respectively, and a greater degree of accuracy at 88.8% and AUC 0.93 by ensembles bagging tree at 5 folds cross validation.

Then the classifier cross validation value was changed to 10 in order to obtain a greater degree of accuracy. At 10 folds cross validation, via the same improved selection feature, the classification results in shown Figure 5.33 were: accuracy 91.3% and AUC 0.92 by ensemble RUSBOOSTED tree classifier.

The shape features classification error rate was 8.7%, because the OD and OC segmentation accuracy suffered due to the main blood vessel and large size of PPA, which affected the disc boundary and cup segmentation.

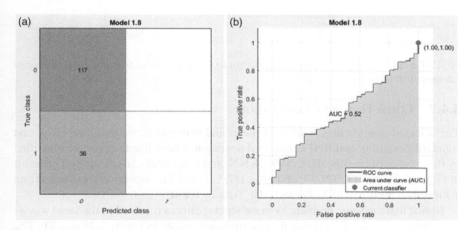

FIGURE 5.31 (a) Shows confusion matrix (b) ROC curve results from the shape selected features.

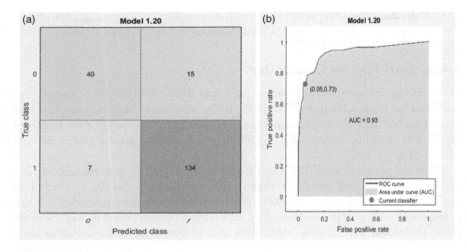

FIGURE 5.32   (a) Shows confusion matrix (b) ROC curve results from the shape balanced features at 5 folds cross validation.

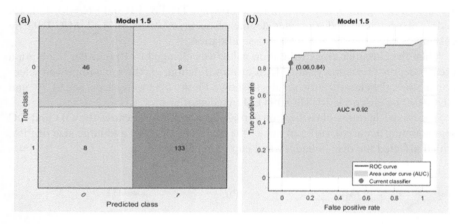

FIGURE 5.33   (a) Shows confusion matrix (b) ROC curve results from the shape balanced features at 10 folds cross validation.

5.4.2 COLOR FEATURES CLASSIFICATION

The selected five features were: cup mean and standard deviation, disc mean and standard deviation, and RNFL standard deviation. These five features are classified by three types of classification: SVM, KNN, and ensembles classification. The result in Figure 5.34 was TP, TN, FP and FN = 117, 1, 1, and 34, respectively, evaluated and achieved an accuracy of 77.1% and AUC 0.68 by SVM.

Notice that the same problem as in the shape features is solved in the same way as in the previous section, and that the classification results in Figure 5.35 are: TP, TN, FP and FN = 57, 133, 17 and 8, respectively. The balanced feature had an accuracy of 88.4% and AUC 0.94 at 5 folds cross validation by ensembles subspace KNN.

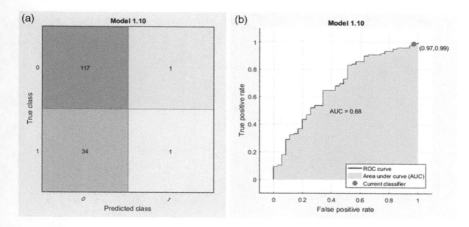

FIGURE 5.34   (a) Shows confusion matrix (b) ROC curve results from the color selected features at 5 folds cross validation.

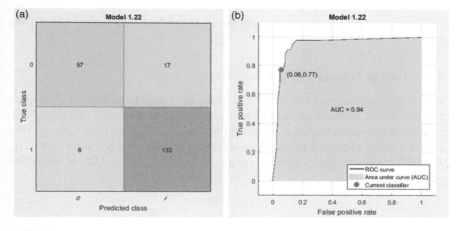

FIGURE 5.35   (a) Shows confusion matrix (b) ROC curve results from the color balanced features at 5 folds cross validation.

Then the classifier cross validation value was changed to 10 to obtain a greater degree of accuracy. At 10 folds cross validation, shown in Figure 5.36: TP, TN, FP and FN = 65, 136, 9, and 5, respectively, the accuracy was 93.5% and AUC 0.96 by ensembles subspace KNN.

The color features classification error rate was 6.5% due to non-uniform illumination happening in the peripheral part of the retina, which often appears darker than the central region because of the curved retinal surface and the geometrical configuration of the light source and camera. These interferences affect the illumination of the ONH and would have an influence on the subsequent statistical analysis in captured fundus images (Abir Ghosh *et al.*, 2015), which used the grid color technique to detect glaucoma and obtained accuracy 87.47% there is a good

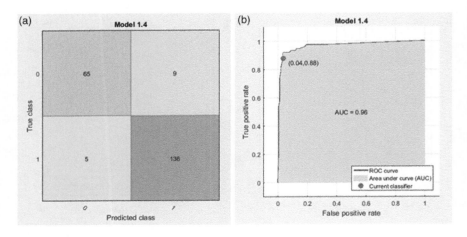

FIGURE 5.36   (a) Shows confusion matrix (b) ROC curve results from the color balanced features at 10 folds cross validation.

improvement in accuracy resulting from the application of color features to specific regions (OD, OC).

5.4.3   TEXTURE FEATURES CLASSIFICATION

The 25 features extracted from the RNFL, using the GLCM and Tamura algorithm, are then selected by the sequential feature selector (SFS) discussed in Chapter 2 to get the most relevant feature the coarseness is finally entered into this feature to the classifier shown in Figure 5.37 and obtained the following: TP, TN, FP and FN = 116, 0, 0, and 37, respectively, with an accuracy of 75.8% and 0.44 AUC by the SVM classifier.

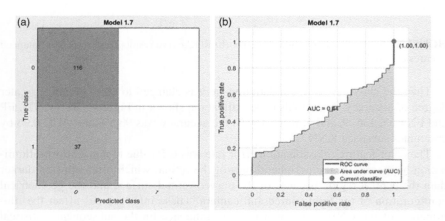

FIGURE 5.37   (a) Shows selected features confusion matrix (b) ROC curve classification results from the texture at 5 folds cross validation.

After balancing the features by the smote algorithm, the classification results in Figure 5.38 improved: TP, TN, FP and FN=54, 128, 10, and 17, respectively, with an accuracy of 87.1% and AUC 0.90 by ensembles boosted tree, ensembles subspace KNN, and ensembles RUSBoosted tree at 5 fold cross validation.

Then the classifier cross validation value was changed to 10 to obtain a greater degree of accuracy. At the 10 fold cross validation, shown in Figure 5.39, the results were: TP, TN, FP and FN=53, 134, 11, and 11, respectively, accuracy was 89.5% and the AUC 0.93 by ensembles RUSBoosted.

The texture features classification errors equal 10.5, and can be explained in the following way. In advanced glaucoma or optic atrophy, RNFL defects cannot be

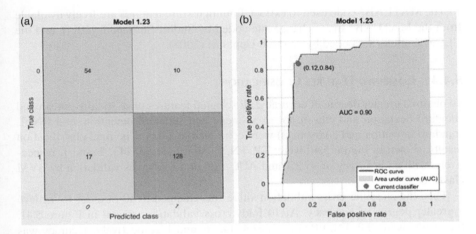

FIGURE 5.38   (a) Shows confusion matrix (b) ROC curve results from the texture balanced features at 5 folds cross validation.

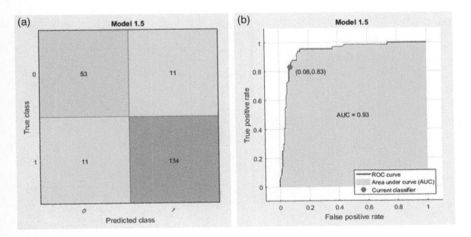

FIGURE 5.39   (a) Shows confusion matrix (b) ROC curve results from the texture balanced features at 10 folds cross validation.

detected since the mean intensity of the RNFL is low in all directions and a localized lesion is not distinguishable. However, in these cases, the pathologic features of disc cupping or atrophy are clearer than RNFL defects and can be easily detected, but, compared with Morris *et al.* (2015), who used the Binary Robust Independent Elementary Features (BRIEF) as the texture features to detect glaucoma and achieved AUC 0.84, and Maya *et al.* (2014), who detected glaucoma by local texture features and achieved a 95.1% success rate with a specificity of 92.3% and a sensitivity of 96.4%, which is better than the texture features presented by this research duo for 2 reasons: the database used was small and the texture features were extracted from the whole image, therefore, the suggested method should work better in detecting glaucoma.

The MATLAB® code used to extract Tamura texture features are freely available from the MATLAB® file exchange site coded by Sudhir Sornapudi (2014), and the GLCM features are coded by Avinash Uppuluri (2008).

5.4.4 COMBINED FEATURES CLASSIFICATION

To obtain a greater degree of accuracy, all selected features (disc sodality, mean and standard deviation and cup minor axes, mean and standard deviation, and RNFL standard deviation and coarseness) were combined together. The final classification results, shown in Figure 5.40 were: TP, TN, FP and FN = 65, 143, 5, and 1, respectively, with an accuracy of 97.2% and AUC 1.00 at 5 fold cross validation by SVM classifier.

Then the classifier cross validation value was changed to 10 in order to achieve a greater degree of accuracy. At 10 folds cross validation, shown in Figure 5.41, the results were: TP, TN, FP and FN = 64, 143, 6, and 1, respectively, accuracy was 96.7% and the ROC curve 1.00 by SVM classifier, where the SVM used before at (Dharmanna *et al.*, 2014), (Guerre et *al.*, 2014) and (Morris T. *et al.*, 2015).

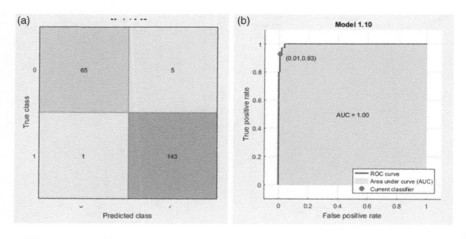

FIGURE 5.40   (a) Shows confusion matrix (b) ROC curve results from the combined features at 5 folds cross validation.

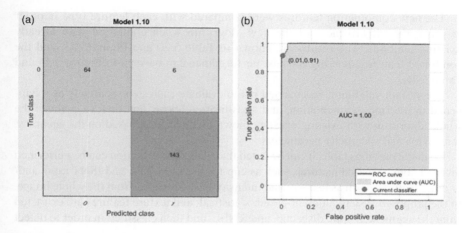

FIGURE 5.41   (a) Shows confusion matrix (b) ROC curve results from the combined features applied to RIM-ONE (version two) at 10 folds cross validation.

To evaluate the performance of the proposed algorithm, the features were tested in a second database named DRISHTI-GS, which contained 70 glaucomatous images and 31 healthy images. The results, shown in Figure 5.42, were as follows: TP, TN, FP and FN = 3, 116, 11, and 0, respectively; accuracy after features balanced at 10 fold cross validation was 91.5% and AUC 0.84 by ensembles Subspace KNN.

The features were tested in a third database named RIM-ONE (version one), which contained 200 glaucomatous images and 255 healthy images. The output results, shown in Figure 5.44, were as follows: TP, TN, FP and FN = 36, 750, 45, and 0, respectively; accuracy after features balanced at 10 fold cross validation was 94.5% and AUC 0.96 by ensembles Subspace KNN.

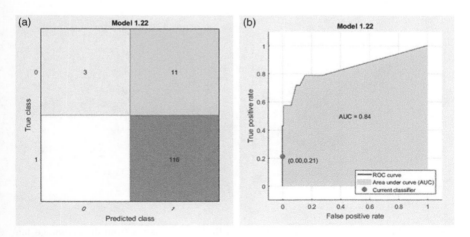

FIGURE 5.42   (a) Shows confusion matrix (b) ROC curve results from the combined features applied to DRISHTI-GS database at 10 folds cross validation.

The new combination features were compared with each feature type in order to prove the effectiveness of the new alogrithm when compared with already existing methods. The results are shown in Table 5.20 and Figure 5.43, and the combined features showing the best performance can be seen in Figure 5.44 and Table 5.21.

The system validation was carried out to evaluate each step separately as it happened in filtering, segmentation, and classification above, and then to evaluate the whole algorithm with existing ones, as shown in Table 5.22, based on the accuracy, sensitivity, and specificity parameters.

From the previous table, it can be noted that glaucoma detection can be performed through domain-based features, such as cup-to-disc ratio, PPA and ISNT ratio, and/ or other features, such as texture- and intensity-based features from the whole image. In the proposed system, domain-based, statistical, and texture features are extracted from the segmented optic disc, cup, and RNFL, and then classified in order to detect glaucoma.

TABLE 5.20
Shows the Different Types of Feature Classification Results at 10 folds

Type of features	Accuracy %	AUC	Classifier
Shape features	91.3	0.92	Ensembles RUSBoosted tree
Color features	93.5	0.96	Ensembles subspace KNN
Texture features	89.5	0.93	Ensembles RUSBoosted tree
Combined features (proposed)	**96.7**	**0.99**	**SVM classifier**

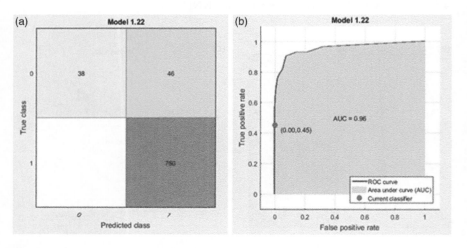

FIGURE 5.43 (a) Shows confusion matrix (b) ROC curve results from the combined features applied to RIM-ONE (version one) at 10 folds cross validation.

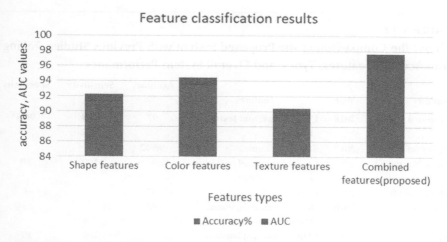

FIGURE 5.44 Bar chart illustrates the different features applied for glaucoma detection.

TABLE 5.21
Shows the Proposed Method Classification Results in the 4 Databases

Database	Accuracy %	AUC
RIM_ONE (version two)	96.7	1.00
DRISHTI_GS	91.5	0.84
RIM_ONE (version one)	94.5	0.96

Compared with Anindita Septiarini *et al.* (2018), who achieved an accuracy of 95.24%, there is a good deal of improvement in using the proposed features. The result is reasonable because of the following considerations:

- The final accuracy is high after merge features (more than each features type separately).
- The classification done by many classifiers obtained a similarly high accuracy and AUC.
- No significant over fitting occurred because the final classification tested at 5 fold cross-validation and 10 fold cross-validation.

The results show the efficiency of the system based on the feature selected by the t-test and SFS method. Next, they show that the detection of glaucoma is based on the final feature selected. Table 5.23 shows the values of the final proposed system's evaluation parameters. The performance of the research method

TABLE 5.22

Shows the Comparison of the Proposed System with Previous Studies, Taking into Account Features, Types, and Overall System Performance

Authors	Year	Features	Accuracy %	Sensitivity %	Specificity %
Proposed system	**2018**	**Color, shape, and texture features**	**97**	**98.4**	**96.6**
R. Geetha Ramani	2017	Statistical features	96.42	–	–
Mohd Nasiruddin	2017	CDR, blood vessel ratio, disc to center distance	–	100	80
Sharanagouda	2017	CDR + ISNT	97	–	–
Claro M.	2016	Disc segmentation, texture feature	93	–	–
Salem	2016	CDR, texture and intensity based features	–	100	87
Swapna'	2016	Fractal Dimension + LBP	88.7	87.2	90
Oh, Yang	2015	RNFL defects	94	86	75
Abir	2015	Grid Color Moment	87.5	–	–
Morris T.	2015	BRIEF	78	–	–
Karthikeyan Sakthivel	2015	LBP + Daugman's algorithm	–	95.4	95.4
Iyyanarappan	2014	DWT	95	–	–
Geetha Ramani	2014	color spaces, channel extraction, statistical, histogram, GLCM	86.67	–	–
Guerre, A.,	2014	CDR	89	93	85
Maya	2014	Local Binary Patterns	–	96.4	92.3
Ganesh Babu	2014	CDR + ISNT	96	96.5	93.3
Preethi	2014	CDR	96	–	–
Fauzia	2013	CDR, ISNT	94	–	–
Rama	2012	HOS, TT, and DWT	91.67	90	93.33
Babu	2011	CDR	90	–	–
Muramatsu	2011	PPA	–	73	95

is expressed in terms of sensitivity, specificity, and accuracy. These values are defined as follows:

Sensitivity (SE) is the ratio of glaucomatous images, which were marked and classified as glaucomatous, to all marked images.
$SE = TP/(TP + FN) = TPR$, from this equation it was calculated to be **0.98**.
Specificity (SP) is the ratio of glaucomatous images, which were marked and classified as healthy, to all marked images.
$SP = TN/(FP + TN) = TNR$, from this equation it was calculated to be **0.97**.

TABLE 5.23

Shows the Comparisons of the Proposed System with Previous Studies That Used the Same Databases

Authors	Used Database	Features	Accuracy %	Sensitivity %	Specificity%
Proposed system 2018	DRISHTI-GS RIM-ONE	Color, shape, and texture features	97	98.4	96.6
Cheriguene 2017	RIM-ONE	TWSVM, and three types of features	98.53	94.11	100
Artem 2017	DRISHTI-GS RIM-ONE	U-Net Convolutional Neural Network	92.08	80	62
Karkuzhali 2017	DRISHTI-GS	CDR, ISNT, DOO	100	100	100
Sharanagouda 2017	DRISHTI-GS	CDR + ISNT	96	–	–
Claro M. 2016	DRISHTI-GS RIM-ONE	Disc segmentation, texture feature	93	–	–
Swapna 2016	DRISHTI-GS	Texture Features Fractal Dimension +LBP.	88.70	87.2	90
Arunava 2016	DRISHTI-GS	Segmentation, Image-based Features	74.1	80	65
Ramaswamy 2016	RIM-ONE	A Depth Based Approach	–	83	83
Maila 2016	DRISHTI-GS RIM-ONE	GLCM,Entropy	93.03	–	–
Abir,2015	RIM-ONE	Grid Color Moment	87.5	–	–

From Figure 5.20, we can notice that:

- **True positive (TP)** is the number images detected as glaucoma by an expert and the proposed method.
- **True negative (TN)** is the number of images detected as normal by an expert and the proposed method.
- **False positive (FP)** is the number of images detected as normal by an expert but detected as glaucoma by the proposed method.
- **False negative (FN)** is the number of images detected as glaucoma by an expert but detected as normal by the proposed method.

The values of sensitivity, specificity, and accuracy lie between 0 and 1. Therefore, if the result of the proposed method is accurate, it should be close to 1.

The performance of our proposed method was evaluated by comparing the results of glaucoma detection provided by an expert. Based on the classification results of our method, using eight features, the marker on the plot shows the performance of the currently selected classifier. The marker shows the values of the false positive rate (FPR) and the true positive rate (TPR) for the currently selected classifier.

For further validation, the proposed algorithm was compared with other glaucoma detection methods that used the same databases (RIM-ONE, DRISHTI-GS). The results are shown in Table 5.23.

From Table 5.23 we noticed that Karkuzhali *et al.* (2017) achieved 100% accuracy, sensitivity, and specificity using a small database of 13 healthy images and 13 glaucomatous images. And Cheriguene *et al.* (2017) obtained an accuracy of 98.5% using a twin support vector machine (TWSVM), where the TWSVM solves a pair of non-parallel hyper-planes, whereas the SVM solves a single complex one. Swapna *et al.* (2016) obtained an accuracy of 88.7% in detecting glaucoma via texture features, which is close to the results obtained here using the texture features alone (Table 5.24).

For both glaucoma detection and classification, our method achieves excellent quality comparable to the following researchers: Claro M. *et al.* (2016), who achieved an accuracy of 93% using texture and OD segmentation; Fauzia Khan *et al.* (2013), who achieved an overall accuracy of 94% using CDR and ISNT rule features; Mohd Nasiruddin *et al.* (2017), who achieved a sensitivity and specificity of 100% and 80%, respectively, using ONH features, outperforming them in terms of the accuracy and type of features; and even Salem *et al.* (2016), who used a combination of structural (cup-to-disc ratio) and non-structural (texture and intensity) measures,

TABLE 5.24

Shows the Final Proposed System's Evaluation Parameters and Values

Sensitivity	Specificity	Overall accuracy
98.4%	96.6%	97%

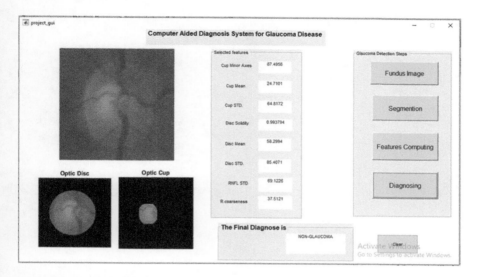

FIGURE 5.45   Glaucoma CAD system.

and achieved an average sensitivity and specificity of 100 and 87%, respectively; and Sakthivel *et al.* (2014) who obtained an accuracy 95% using GLCM features. Other algorithms based on combined features were used in the study presented by Sirel *et al.* (2017), which concluded that the RNFL defects in OCT images alone may be misleading in glaucoma examination, and that it needs another type of feature to get an accurate diagnosis, whereas Dharmanna *et al.* (2014) found that there was a linear correlation coefficient between the RNFL defects and the texture and fractal dimensions. Therefore, the CAD system could be used in a low-priced medical screening setting, thus avoiding the inter-experts variability issue by introducing the segmentation, features, and final diagnosing to its users, as shown in Figure 5.45.

6 Conclusion and Future Scope

6.1 CONCLUSIONS

In conclusion, glaucoma is a group of eye diseases that have no symptoms and, if not detected at an early stage, may cause permanent blindness; some preceding structural damage to the retina is one of the marked symptoms of glaucoma. It is diagnosed by an examination of the size, structure, shape, and color of the optic disc, optic cup, and retinal nerve fiber layer (RNFL), and, due to the subjectivity of human experience, fatigue factor, etc., there is a need for a CAD system to manage large volumes of data and provide objective assessments for decision support and help in labor-intensive, observer-driven tasks using the fundus images, which, is among one of the main biomedical imaging techniques to analyze the internal structure of the retina. The proposed technique provides a novel algorithm to detect glaucoma from a digital fundus image. It uses MATLAB® software evaluated on a RIM_ONE (version two) database, containing digital fundus images from 158 patients (118 healthy images and 40 glaucomatous images) and DRISHTI_GS, which contains 101 digital fundus images (70 glaucomatous images and 31 healthy images), and RIM-ONE (version one) (200 healthy images and 250 glaucomatous images), where the proposed approach used to detect glaucoma was carried out via three steps: firstly, OD and OC segmentation. In OD and OC segmentation several steps were done like pre-processing, thresholding, boundary smoothing, and disc reconstruction to a full circle, where OD segmentation achieved a best dice coefficient (DSC) of 90% and a Structural Similarity (SSIM) of 83%, and OC segmentation results were a dice coefficient of 73% and a Structural Similarity (SSIM) of 93%, and cup segmentation achieved an SSIM of 93%; secondly, shape, color, and texture features were extracted from the segmented parts and then the most relevant features were selected; thirdly, many types of classifier were applied to find the best classification accuracy via a set of color-based, shape-based, and texture features by extracting 13 shape features from disc and cup, and extracting 25 texture features from RNFL (retinal nerve fiber layer) using the gray level co-occurrence method, Tamar algorithm, and 3 color features for each of disc, cup, and RNFL. Next, best features were selected by T-test method and Sequential feature selection (SFS) to introduce eight features (cup minor axes, disc and cup mean, standard deviation, and RNFL standard deviation and coarseness) with an average accuracy of 97%, maximizing the area under the curve (AUC) 0.99 using the SVM classifier. The key contribution in this work proposes new features that are suitable for glaucoma detection.

In this study, an efficient approach for building a computerized system for glaucoma detection using digital fundus images is presented and discussed. The automated combined features system is a reliable and efficient method for glaucoma diagnosis. This research work focuses the digital fundus image analysis for glaucoma assessment to develop an automatic glaucoma detection system based on the evaluation of many features.

It is observed from the experimental results that the proposed system achieves better accuracy for glaucoma with combined features than each type of features individually, and the proposed system achieves very promising results with an accuracy of 97%, a specificity of 96.6%, and a sensitivity of 98.4%. The objectives of this research are all met. The early detection system for glaucoma diagnosis is effectively implemented, and features are computed to classify whether the corresponding digital fundus image is healthy or subjected to glaucoma. The experimental results pointed out that the use of shape, color, and texture features of correctly segmented OC, OD, and RNFL regions are affected in glaucoma diagnosis.

The proposed CAD system will certainly assist ophthalmologists in attaining a better diagnosis of glaucoma, and should make a valuable contribution to medical science by supporting medical image analysis for glaucoma detection.

6.2   FUTURE SCOPE

As the field of interest and the results of this study turned out to be rich and broad, there are several ways to extend it. Some of the possible ways to investigate this work in the near future are discussed below:

1. Design a database connected with the software to save the patients' fundus images and medical reports.
2. Design a complete, integrated, automated system to classify all of the different types of glaucoma, namely: Primary Open-Angle Glaucoma, Normal Tension Glaucoma, Angle Closure Glaucoma, Acute Glaucoma, Exfoliation Syndrome, and Trauma-Related Glaucoma.
3. Complete the system to not just diagnose glaucoma but compute the progress of the disease by comparing the different images of the same patient to be used for follow-up.
4. Design a holder to modify smartphones to be simple and cheap fundus camera devices with a wide range of availability.

References

Abdullah, M., M.M. Fraz, and S.A. Barman (2016). Localization and segmentation of optic disc in retinal images using circular Hough trans- form and grow-cut algorithm, *PeerJ*, 4: e2003.

Abràmoff, M.D., M.K. Garvin, and M. Sonka (2010). Retinal imaging and image analysis, *IEEE Reviews in Biomedical Engineering*, 3: 169–208.

Achanta, R., A. Shaji, K. Smith, A. Lucchi, and S. Süsstrunk (2012). Slic super pixels compared to state-of-the-art super pixel methods, *IEEE Transactions on Pattern Analysis and Machine Intelligence*, 34(11): 2274–2281.

Acharya, U.R., S. Dua, X. Du, S.V. Sree, and C.K. Chua (2011). Automated diagnosis of glaucoma using texture and higher order spectra features, *IEEE Transactions on Information Technology in Biomedicine*, 15(3): 449–455.

Ahmed, Arwa, and Alnazier Osman (2018). Optic disc segmentation using manual thresholding technique, *Journal of Clinical Engineering*, 44(1): 28–34. January/March 2019, Copyright © 2018 Wolters Kluwer Health.

Alghamdi, M., M. Al-Mallah, S. Keteyian, C. Brawner, J. Ehrman, and S. Sakr (2017). Predicting diabetes mellitus using SMOTE and ensemble machine learning approach: The Henry Ford ExercIse Testing (FIT) project, *PLoS One*, 12(7): e0179805.

Allam, Ali (2017). Automatic detection of landmarks and abnormalities in eye fundus images (February), publication at: www.researchgate.net/publication/308611012. doi:10.13140/ RG.2.2.23488.12800.

Allingham, R.R., and K.F. Damji (2005). *Shields Textbook of Glaucoma*, 5th ed. Lippincott Williams and Wilkins, Philadephia, PA.

Almazroa, Ahmed, Ritambhar Burman, Kaamran Raahemifar, and Vasudevan Lakshminarayanan (2015). Optic disc and optic cup segmentation methodologies for glaucoma image detection: A survey, *Journal of Ophthalmology*, Article ID 180972, 28 pages. doi:10.1155/2015/180972.

Almazroa, Ahmed, Weiwei Sun, Sami Alodhayb, Kaamran Raahemifar and Vasudevan Lakshminarayanan (2017). Optic disc segmentation for glaucoma screening system using fundus images, *Clinical Ophthalmology*, 11: 2017–2029.

Almazroa, Ahmed, Weiwei Sun, Sami Alodhayb, Kaamran Raahemifar, and Vasudevan Lakshminarayanan (2017). Optic disc segmentation: Level set methods and blood vessels inpainting, 1013806. doi:10.1117/12.2254174.

Alward, W.L. (1998). Medical management of glaucoma, *New England Journal of Medicine*, 339(18): 1298–1307.

Aslam, Javed A., Raluca A. Popa, and Ronald L. Rivest (2007). On estimating the size and confidence of a statistical audit, Proceedings of the Electronic Voting Technology Workshop, Boston, Massachusetts, (August 6, 2007).

Bandodkar, P., R. Birdwell, and D. Ikeda (2002). Computer aided detection (CAD) with screening mammography in an academic institution: Preliminary findings, *Radiology*, 225: 458.

Bermingham, Mairead L., R. Pong-Wong, A. Spiliopoulou, *et al.* (2015). Application of high-dimensional feature selection: Evaluation for genomic prediction in man, *Scientific Reports*, 5: 10312.

Bernardes, R., P. Serranho, and C. Lobo (2011). Digital ocular fundus imaging: A review, *Ophthalmological*, 226(4): 161–181.

Blackledge, Jonathan M. (2005). *Digital Image Processing,* 1st edition *Woodhead Publishing, Cambridge, Massachusetts.*

Bock, Rüdiger, Jörg Meier, László G. Nyúl, Joachim Hornegger, and Georg Michelson (2010). Glaucoma risk index: Automated glaucoma detection from color fundus images, *Medical Image Analysis*, 14(3): 471–481.

Booysen, Dirk (2013). A review of fundus autofluorescence imaging, *South Afircan Optometrist*, 72(1). doi:10.4102/aveh.v72i1.38.

Bourne, R.R.A., H.R. Taylor, S.R. Flaxman, *et al.* (2016). Number of people blind or visually impaired by glaucoma worldwide and in world regions 1990–2010: A meta-analysis, *PLoS One*, 11(10): e0162229.

Bowd, C., L.M. Zangwill, C.C. Berry, E.Z. Blumenthal, C. Vasile,C. Sanchez-Galeana, C.F. Bosworth, P.A. Sample, and R.N. Weinreb (2001). Detecting early glaucoma by assessment of retinal nerve fiber layerthickness and visual function, *Investigative Ophthalmology and Visual Science*, 42(9): 1993–2003.

Brady, A., R.Ó. Laoide, P. McCarthy, and R. McDermott (2012). Discrepancy and error in radiology concepts and causes and consequences, *Ulster Medical Journal*, 81(1): 3–9.

Breiman, Leo (1996). Bagging predictors, *Machine Learning*, CiteSeerX 10.1.1.32.9399, 24(2): 123–140.

Burroni, Marco, Rosamaria Corona, Giordana Dell'Eva, Francesco Sera, Riccardo Bono, Pietro Puddu, Roberto Perotti, Franco Nobile, Lucio Andreassi, and Pietro Rubegni (2004). Melanoma computer-aided diagnosis: Reliability and feasibility study, *Clinical Cancer Research*, 10: 1881–1886.

Cello, K.E., J.M. Nelson-Quigg, and C.A. Johnson (2000). Frequency doubling technology perimetry for detection of glaucomatous visual field loss, *American Journal of Ophthalmology*, 129(3): 314–322.

Chan, T.F., and L.A. Vese (2001). Active contours without edges, *IEEE Transactions on Image Processing*, 10(2): 266–277.

Chakravarty, Arunava, and Jayanthi Sivaswamy (2016). Glaucoma classification with a fusion of segmentation and image-based features, 978-1-4799-2349-6/16/$31.00 ©2016 IEEE.

Chaudhary, A., and T. Gulati (2013). Segmenting digital images using edge detection, *International Journal of Application or Innovation in Engineering & Management (IJAIEM)*, 2(5).

Chawla, Nitesh V., Kevin W. Bowyer, Lawrence O. Hall, and W. Philip Kegelmeyer (2002). Smote: Synthetic minority over-sampling technique, *Journal of Artificial Intelligence Research (JAIR)*, 16: 321–357.

Chen, C.-M., Chou, Yi-Hong, Norio Tagawa, and Younghae Do (2013). Computer-aided detection and diagnosis in medical imaging, *Computational and Mathematical Methods in Medicine*, 2013. Article ID 790608, doi:10.1155/2013/790608.

Cheng, Jun, Jiang Liu, Yanwu Xu, *et al.* (2013). Superpixel classification based optic disc and optic cup segmentation for glaucoma screening, *IEEE Transactions on Medical Imaging*, 32(6): 1019–1032.

Cheriguene, S. N. Azizi, H. Djellali, O. Bounakhla, M. Aldwairi, and A. Ziani (2017). New computer aided diagnosis system for glaucoma disease based on twin support vector machine, First international conference on Embedded & Distributed Systems, EDiS.

Chrastek, R., M. Wolf, K. Donath, G. Michelson, and H. Niemann (2002). Optic disc segmentation in retinal images, Proceedings of Bildverarbeitung für die Medizin, Leipzig, Germany, March 2002, pp. 263–266.

Claro, M., Leonardo Santos, Wallinson Silva, Flávio Araújo, Nayara Moura, and André Macedo (2016). Automatic glaucoma detection based on optic disc segmentation and texture feature extraction, *CLEI Electronic Journal*, 19(2): 4.

Claro, Maila, Leonardo Santos, Wallinson Silva, Fliavio Araiujo, Nayara Moura, and André Macedo (2016). Automatic glaucoma detection based on optic disc segmentation and texture feature extraction, *Clei Electronic Journal*, 19(2): 4.

Coleman, A.L., and S. Miglior (2008). Risk factors for glaucoma onset and progression, *Ophthalmology*, 53(6): S3–S10.

Coste, Arthur. Histograms, University of Utah: CS6640 image processing report, www.sci.utah.edu/~acoste/uou/Image/project1/Arthur_COSTE_Project_1_report.html.

Damms, T., and F. Dannheim (1993). Sensitivity and specificity of optic disc parameters in chronic glaucoma, *Investigative Ophthalmology & Visual Science*, 34(7): 2246–2250.

Daniilidis, K., P. Maragos, and N. Paragios (Eds.) (2010). ECCV 2010, Part I, LNCS 6311, 1–14, 2010. c Springer-Verlag Berlin Heidelberg 2010.

Deepikaa, L., Mary Gladenceb, and S. Kalpanac (2016). Detection of glaucoma based on optic disc and optic cup segmentation using slice super pixels, *International Journal of Pharmacy & Technology*, 8(4): 22781–22792.

Devasia, Thresiamma, Poulose Jacob, and Tessamma Thomas (2015). Fuzzy clustering based glaucoma detection using the CDR, *Signal & Image Processing: An International Journal (SIPIJ)*, 6(3), 55–70.

Dhawan, A., and S. Dai (2008). Clustering and pattern classification. In: *Principles and Advanced Methods in Medical Imaging and Image Analysis*. Singapore: World Scientific Publishing Co. Pte. Ltd., pp. 229–265.

Doi, K. (2017). Computer-aided diagnosis in medical imaging: Historical review, current status and future potential. *Computerized Medical Imaging and Graphics*, 31(4): 198–211.

Dougherty, G. (2009). *Digital Image Processing for Medical Applications*. New York: Cambridge University Press. ISBN-13 978-0-511-53343-3.

Downs, J.C., M.D. Roberts, and I.A. Sigal (2011). Glaucomatous cupping of the lamina crib Rosa: A review of the evidence for active progressive modeling as a mechanism, *Experimental Eye Research*, 93(2): 133–140.

Drance, S., D.R. Anderson, and M. Schulzer (2001). Risk factors for progression of visual field abnormalities in normal-tension glaucoma, *American Journal of Ophthalmology*, 131(6): 699–708.

Ehrlich, Joshua, and Nathan Radcliffe (2010). The role of clinical parapapillary atrophy evaluation in the diagnosis of open angle glaucoma. *Clinical Ophthalmology (Auckland, N.Z.)*, 4: 971–976. doi:10.2147/OPTH.S12420.

Elseid, Arwa A. Gasm, and Alnazier O. Hamza (2018). Glaucoma detection based on shape features and SMOTE algorithm, *CiiT International Journal of Digital Image Processing*, 10(10), October–November.

Elseid, Arwa Ahmed Gasm, and Mohamed Eltahir Elmanna (2018). Optic cup segmentation using manual thresholding level technique, *International Journal of Current Research in Life Sciences*, 7(10): 2769–2773.

Elseid, Arwa Ahmed Gasm, Mohamed Eltahir Elmanna, and Alnazier Osman Hamza (2018). Evaluation of spatial filtering techniques in retinal fundus images, *American Journal of Artificial Intelligence*, 2(2): 16–21.

Elseid, Arwa Ahmed Gasm, Alnazier Osman Hamza, and Ahmed Fragoon (2018). Developing a real time algorithm for diagnosing glaucoma, *ICTACT Journal on Image and Video Processing*, 9(2), 1894–1900.

Engelhorn, T., G. Michelson, S. Waerntges, T. Struffert, S. Haider, and A. Doerfler (2011). Diffusion tensor imaging detects rarefaction of optic radiation in glaucoma patients. *Academic Radiology*, 18(6): 764–769.

Fechtner, R., and R. Weinreb (1994). Mechanisms of optic nerve damage inprimary open angle glaucoma, *Survey of Ophthalmology*, 39(1): 23–42.

Fernández, Cabrera Delia, Harry M. Salinas, and Carmen A. Puliafito (2006). Automated detection of retinal layer structures on optical coherence tomography images, *Optical Society of America.* doi:10.1364/OPEX.13.010200.

Fernandez-Granero, M.A. A. Sarmiento, D. Sanchez-Morillo, S. Jiménez, P. Alemany, and I. Fondón (2017). Automatic CDR estimation for early glaucoma diagnosis, *Journal of Healthcare Engineering*, 2017. Article ID 5953621, 14 pages, https://doi.org/ 10.1155/2017/ 5953621.

Ferreras, A., L. Pablo, J. Larrosa, V. Polo, A.B. Pajarin, and F.M.Honrubia (2008). Discriminating between normal and glaucoma-damaged eyes with the Heidelberg retina Tomograph 3, *Ophthalmology*, 115(5), 775–781.e2.

Ferri, F.J., P. Pudil, M. Hatef, and J. Kittler (1994). Comparative study of techniques for large-scale feature selection, *Pattern Recognition in Practice*, IV: 403–413.

Foster, P.J., R. Buhrmann, H.A. Quigley, and G.J. Johnson (2002). The definition and classification of glaucoma in prevalence surveys, *British Journal of Ophthalmology*, 86(2): 238–242.

Fumero, F., J. Sigut, and S. Alayón (2015). Interactive tool and database for optic disc and cup segmentation of stereo and monocular retinal fundus images, WSCG 2015 Conference on Computer Graphics, Visualization and Computer Vision, 91–97, ISBN 978-80-86943-66-4.

Fumero, F., S. Alayon, J.L. Sanchez, J. Sigut, and M. Gonzalez-Hernandez (2011). RIM-ONE: An open retinal image database for optic nerve evaluation, Proceedings of the IEEE Symposium on Computer-Based Medical Systems (July 2011). doi:10.1109/ CBMS.2011.5999143

Freer, T.W., and M.J. Ulissey (2001). Screening mammography with computer aided detection: Prospective study of 12,860 patients in a community breast center, *Radiology*, 220: 781–786.

Fu, Huazhu, Jun Cheng, Yanwu Xu, Damon Wing Kee Wong, Jiang Liu, and Xiaochun Cao (2018). Joint optic disc and cup segmentation based on multi-label deep network and polar transformation, arXiv: 1801.00926v3 [cs.CV], 1–9.

Fujitaa, Hiroshi, Yoshikazu Uchiyamaa, Toshiaki Nakagawaa, *et al.* (2008). Computer-aided diagnosis: The emerging of three CAD systems induced by Japanese health care needs, *Computer Methods and Programs in Biomedicine*, 92: 238–248.

Gandhi, Monica, and Suneeta Dubey (2013). Evaluation of the optic nerve head in glaucoma, *Journal of Current Glaucoma Practice*, 7(3): 106–114.

Ganesh Babu, T.R., and S. Shenbagadevi (2011). Automatic detection glaucoma using fundus image, *European Journal of Scientist Research*, 59(1): 22–32.

Ganesh Babu, T.R., R. Sathishkumar, and Rengara Jvenkatesh (2014), Segmentation of optic nerve head for glaucoma detection using fundus images, *Biomedical & Pharmacology Journal*, 7(2), 697–705.

Gao, Yue, and Qionghai Dai (2015). *View-Based 3-d Object Retrieval.*

Garaci, F., F. Bolacchi, A. Cerulli, M. Melis, A. Spano, C. Cedrone,R. Floris, G. Simonetti, and C. Nucci (2009). Optic nerve and optic radiation neuro degeneration in patients with glaucoma: In vivo analysis with 3-Tdiffusion-tensor MR imaging, *Radiology*, 252(2): 496–501.

Geetha Ramani, R., C. Dhanapackiam, and B. Lakshmi (2014). Automatic detection of glaucoma in fundus images through image features, International Conference on Knowledge Modelling and Knowledge Management, pp. 135–144. ISBN -9789351377658

Geetha Ramani, Sugirtharani S., and B. Lakshmi (2017). Automatic detection of glaucoma in retinal fundus images through image processing and data mining techniques, *International Journal of Computer Applications* (0975-8887), 166(8), 38–43.

Ghosh, A., A.S. Ashour, N. Dey, A. Sarkar, D. Bălas-Timar, and V.E. Balas (2015). Grid color moment features in glaucoma classification, *(IJACSA) International Journal of Advanced Computer Science and Applications*, 6(9): 1–4.

Giger, M.L. (2000). Computer-aided diagnosis in mammography. In: *Handbook of Medical Imaging*, 2nd ed. SPIE Digital Library, Europe, pp. 915–1004.

Giger, M.L. (2010). Department of Radiology, University of Chicago, Chicago, Illinois.

Gómez-Valverde, Juan J., Alfonso Antón, Gianluca Fatti, Bart Liefers, Alejandra Herranz, Andrés Santos, Clara I. Sánchez, and María J. Ledesma-Carbayo (2019). Automatic glaucoma classification using color fundus images based on convolutional neural networks and transfer learning, *Biomedical Optics Express*, 10(2): 892–913.

Godoy, M.C., T.J. Kim, C.S. White, *et al.* (2013). Benefit of computer-aided detection analysis for the detection of subsolid and solid lung nodules on thin- and thick-section CT, *American Journal of Roentgenology*, 200(1): 74–83. doi:10.2214/AJR.11.7532.

Gonzalez, R.C., and R.E.Woods (2008). *Digital Image Processing*, 3rd ed, Pearson International Edition prepared by Pearson Education.

Grady, Denise (1993). The vision thing: Mainly in the brain, Discover (June 01, 1993), http:// discovermagazine.com /1993/jun/ thevisionthingma227.

Guerre, A., J. Martinez-del-Rincon, P. Miller, and A. Azuara-Blanco (2014). Automatic analysis of digital retinal, images for glaucoma detection, Paper presented at Irish Machine Vision and Image Processing Conference, Derry, United Kingdom.

Güngör, Sirel Gür, and Ahmet Akman (2017). Are all retinal nerve fiber layer defects on optic coherence tomography glaucomatous? *The Turkish Journal of Ophthalmology*, 47: 267–273.

Gupta, N., and Y. Yucel (2007). What changes can we expect in the brain of glaucoma patients? *Survey of Ophthalmology*, 52(6): S122–S126.

Gupta, N., L. Ang, L. de Tilly, L. Bidaisee, and Y. Yucel (2006). Human glaucoma and neural degeneration in intracranial optic nerve, lateral geniculate nucleus, and visual cortex, *British Journal of Ophthalmology*, 90(6): 674–678.

Guyon, Isabelle, and André Elisseeff (2003). An introduction to variable and feature selection, *Journal of Machine Learning Research*, 3: 1157–1182.

Haddock, J. Luis, and Allie Nadelson (2016). Teleophthalmology of retinal diseases, *Retinal Physicians*, 13(April): 34, 36–38.

Haleem, Muhammad Salman, Liangxiu Han, Jano van Hemert, and Baihua Li (2013). Automatic extraction of retinal features from colour retinal images for glaucoma diagnosis: A review, *Computerized Medical Imaging and Graphics*, 37: 581–596. doi:10.1016/j.compmedimag.2013.09.005.

Harizman, N., C. Oliveira, A. Chiang, C. Tello, M. Marmor, R. Ritch, and J.M. Liebmann (2006). The ISNT rule and differentiation of normal, from glaucomatous eyes. *Archives of Ophthalmology*, 124: 1579–1583.

He, Haibo, Yang Bai, Edwardo A. Garcia, and Shutao Li (2008). ADASYN: Adaptive synthetic sampling approach for imbalanced learning, Proceedings of the International Joint Conference on Neural Networks, pp. 1322–1328. doi:10.1109/IJCNN.2008.4633969.

Hussain, A.R. (2008). Optic nerve head segmentation using genetic active contours, International, Conference on Computer and Communication Engineering, ICCCE 2008, (13–15 May 2008), pp. 783–787, Kuala Lumpur, Malaysia.

Hussain, Syed Akhter (2015). Automated detection and classification of glaucoma from eye fundus images: A survey, *International Journal of Computer Science and Information Technologies*, 6(2), 1217–1224.

Ingle, R., and P. Mishra (2013). Cup segmentation by gradient method for the assessment of glaucoma from retinal image, *International Journal of Engineering Trends and Technology*, 4(6): 2540–2543.

Irshad, S., Xiaoxia Yin, Lucy Qing Li, and Umer Salman (2016). Automatic optic disk segmentation in presence of disk blurring, springer international publishing, part I, *LNCS*, 10072: 13–23.

Iyyanarappan, A., and G. Tamilpavai (2014). Glaucomatous image classification using wavelet based energy features and PNN, *International Journal of Technology Enhancements and Emerging Engineering Research*, 2(4): 85–90. ISSN: 2347-4289.

Jacob, Eleesa, and R. Venkatesh (2014). A method of segmentation for glaucoma screening using superpixel classification, *International Journal of Innovative Research in Computer and Communication Engineering* (An ISO 3297: 2007 Certified Organization), 2(1), 2536–2544.

James, Gareth, Daniela Witten, Trevor Hastie, and Robert Tibshirani (2013). *An Introduction to Statistical Learning*. Springer, p. 204.

Jan, J., J. Odstrcilik, J. Gazarek, and R. Kolar (2012). Retinal image analysis aimed at blood vessel tree segmentation and early detection of neural-layer deterioration, *Computerized Medical Imaging and Graphics*, 36(6): 431–441.

Jaskowiak, P.A., and R.J.G.B. Campello (2011). Comparing correlation coefficients as dissimilarity measures for cancer classification in gene expression data, Research Gate Conference, publication at: www.researchgate.net/publication/260333185.

Jonas, J.B., M.C. Fernandez, and G.O. Naumann (1992). Glaucomatous parapapillary atrophy: Occurrence and correlations, *Archives of Ophthalmology*, 110(2): 214–222.

Jonas, J.B., M.C. Fernández, and J. Stürmer. (1993). Pattern of glaucomatous neuroretinal rim loss, *Ophthalmology*, 100(1): 63–68.

Jonas, J.B., W.M. Budde, and S. Panda-Jonas (1999). Ophthalmoscopic evaluation of the optic nerve head, *Survey of Ophthalmology*, 43(4): 293–320.

Jonas, Jost, Pascal Weber, Natsuko Nagaoka, and Kyoko Ohno-Matsui (2017). Glaucoma in high myopia and parapapillary delta zone, *PLoS One*, 12: e0175120.

Joshi, G.D., J. Sivaswamy, and S.R. Krishnaas (2011). Optic disk and cup segmentation from monocular color retinal images for glaucoma assessment, *IEEE Transactions on Medical Imaging*, 30(6): 1192–1205.

Joshi, Gopal Datt, Jayanthi Sivaswamy, Kundan Karan, and S.R. Krishnadas (2010). Optic disk and cup boundary detection using regional information, Proceedings of the 2010 IEEE International Conference on Biomedical Imaging: From Nano to Macro, 948–951. 10.1109/ISBI.2010.5490144.

Joshi, G.D., J. Sivaswamy, K. Karan, R. Prashanth, and S.R. Krishnadas (2010). Vessel bend-based cup segmentation in retinal images, 20th International Conference on Pattern Recognition (ICPR), 2010, CVIT, IIIT Hyderabad, Hyderabad, India, (August 23–26, 2010), pp. 2536–2539.

Jung, Na Young, Bong Joo Kang, Hyeon Sook Kim, Eun Suk Cha, Jae Hee Lee, Chang Suk Park, In Young Whang, Sung Hun Kim, Yeong Yi An, and Jae Jeong Choi (2014). Who could benefit the most from using a computer-aided detection system in full-field digital mammography? *World Journal of Surgical Oncology*, 12: 168.

Kajić, V., B. Povazay, B. Hermann, B. Hofer, D. Marshall, P.L. Rosin, and W. Drexler (2010). Robust segmentation of intra retinal layers in the normal human fovea using a novel statistical model based on texture and shape analysis, *Opt Express*, 18(14): 14730–14744.

Kamat, Vaishnavi, Shruti Chatti, Alvira Rodrigues, Chinmayee Shetty, and Anusaya Vadji (2017). Glaucoma detection using enhanced K-strange points clustering algorithm and classification, *IOSR Journal of Computer Engineering (IOSR-JCE)*, 19(4): 44–49. e-ISSN: 2278-0661, p-ISSN: 2278-8727.

Karkuzhali, S., and D. Manimegalai (2017). Computational intelligence-based decision support system for glaucoma detection, *Biomedical Research*, 28(11): 4737–4748.

Karthikeyan, Sakthivel, and N. Rengarajan (2014). Performance analysis of gray level co-occurrence matrix texture features for glaucoma diagnosis, *American Journal of Applied Sciences*, 11: 248–257. 2014 ISSN: 1546-9239 ©2014 Science Publication.

Kass, M., A. Witkin, and D. Terzopoulos (1988). Snakes: Active contour models, *International Journal of Computer Vision*, 1(4): 321–331.

Kavitha, K., and M. Malathi (2014). Optic disc and optic cup segmentation for glaucoma classification, *International Journal of Advanced Research in Computer Science & Technology (IJARCST 2014)*, 2(Issue Special): 87–90.

Khalida, Noor Elaiza Abdul, Noorhayati Mohamed Noora, Norharyati Md. Ariffa (2014). Fuzzy c-means (FCM) for optic cup and disc segmentation with morphological operation, *Procedia Computer Science*, 42: 255–262.

Khan, Fauzia, Shoaib A. Khan, Ubaid Ullah Yasin, Ihtisham ul Haq, and Usman Qamar (2013). Detection of glaucoma using retinal fundus images, The 2013 Biomedical Engineering International Conference (BMEiCON-2013). doi:10.1109/BMEiCon.2013.6687674.

Koprowski, R., M. Rzendkowski, and Z. Wróbel (2014). Automatic method of analysis of OCT images in assessing the severity degree of glaucoma and the visual field loss, *BioMedical Engineering Online*, 13(1): 16.

Koprowski, R., S. Teper, Z. Wrobel, and E. Wylegala (2013). Automatic analysis of selected choroidal diseases in OCT images of the eye fundus. *BioMedical Engineering Online*, 12: 117.

Koprowski, R., and Z. Wróbel (2011). *Image Processing in Optical Coherence Tomography: Using Matlab*. Katowice, Poland: University of Silesia, Copyright © www.ncbi.nlm.nih.gov/books/NBK97169.

Kotecha (2002). Clinical examination of the glaucomatous patient [Online], www.optometry.co.uk.

Kou, W. Chen, C. Wen, and Z. Li (2015). Gradient domain guided image filtering, *IEEE Transactions on Image Processing*, 24(11): 4528–4539.

Krishnan, M. Muthu Rama, and Oliver Faust (2012). Automated glaucoma detection using hybrid feature extraction in retinal fundus images, *Journal of Mechanics in Medicine and Biology*, 13(1): 1350011–1350032.

Kroese, M., and H. Burton (2003). Primary open angle glaucoma: The need for a consensus case definition, *Journal of Epidemiology and Community Health*, 57(9): 752–754.

Kuehn, M., J. Fingert, and Y. Kwon (2005). Retinal ganglion cell death in glaucoma: Mechanisms and neuroprotective strategies, *Ophthalmology Clinics of North America*, 18: 383–395.

Kumar, Gaurav, and Pradeep Kumar Bhatia (2014). A detailed review of feature extraction in image processing systems, Research Gate, Conference Paper, (February 2014). doi:10.1109/ACCT.2014.74.

Kumar Arun, M.N., and H.S. Sheshadri (2012). Building accurate classifier for the classification of micro calcification, *(IJCSIT) International Journal of Computer Science and Information Technologies*, 3(6): 5346–5350.

Kumar, P.S.J., and Sukanya Banerjee (2014). A survey on image processing techniques for glaucoma detection, *International Journal of Advanced Research in Computer Engineering & Technology (IJARCET)*, 3(12): 4066–4073.

Kushwaha, Sumit, and Rabindra Kumar Singh (2015). Study and analysis of various image enhancement method using MATLAB®. *International Journal of Computer Sciences and Engineering (IJCSE)*, 3.

Kwon, Y.H., J.H. Fingert, M.H. Kuehn, and W.L. Alward (2009). Primary open-angle glaucoma, *New England Journal of Medicine*, 360(11): 1113–1124.

Lalonde, M., M. Beaulieu, and L. Gagnon (2001). Fast and robust optic disc detection using pyramidal decomposition and hausdorff-based template matching, *IEEE Transactions on Medical Imaging*, 20(11): 1193–1200.

Lamani, Dharmanna, T.C. Manjunath, M Mahesh, and Y.S. Nijagunarya (2014). Early detection of glaucoma through retinal nerve fiber layer analysis using fractal dimension and texture feature, *International Journal of Research in Engineering and Technology*, 03(10). eISSN: 2319-1163 | pISSN: 2321-7308.

Leach, Richard (2014). *Fundamental Principles of Engineering Nanometrology*, 2nd ed, William Andrew, **ISBN:** 9781455777501.

Li, Huiqi, and Opas Chutatape (2003). Boundary detection of optic disc by a modified ASM method, *The Journal of the Pattern Recognition Society*, 36: 2093–2104.

Li, Q., and R.M. Nishikawa (Ed.) (2015). *Computer-Aided Detection and Diagnosis in Medical Imaging*. New York: Taylor & Francis, CRC Press.

Lim, Ridia, Ivan Goldberg, Franzco Fracs, Paul N. Schacknow, and John R. Samples (2010). *The Glaucoma Book: A Practical Evidence-Based Approach to Patient Care*. New York: Springer-Verlag (Chapter 1, p. 3).

Lim, T.C., S. Chattopadhyay, and U.R. Acharya (2012). A survey and comparative study on the instruments for glaucoma detection, *Journal of Biomedical Engineering and Technology*, 34(2): 129–139.

Ling, Charles X., and Victor S. Sheng (2008). Cost-sensitive learning and the class imbalance problem. In: C. Sammut (Ed.) *Encyclopedia of Machine Learning*. Springer.

Lu, Shijian, and Joo Hwee Lim (2010). Automatic optic disc detection through background estimation, Proceedings of 2010 IEEE 17th International Conference on Image Processing, Hong Kong, (September 26–29, 2010).

Lusch, David P. (2015). GEO 827, Digital Image Processing and Analysis. October 2015.

Mahalakshmi, V., and S. Karthikeyan (2014). Clustering based optic disc and optic cup segmentation for glaucoma detection, *International Journal of Innovative Research in Computer and Communication Engineering*, 2(4): 3756–3761, An ISO 3297: 2007 Certified Organization.

Malakar, Annesha, and Joydeep Mukherjee (2013). Image clustering using color moments, histogram, edge and K-means clustering, *International Journal of Scien and Research*, 2(1): 532–537. India Online ISSN: 2319–7064.

Maldhure, Prasad N., and V.V. Dixit (2015). Glaucoma detection using optic cup and optic disc segmentation, *International Journal of Engineering Trends and Technology (IJETT)*, 20(2): 52–55.

Manohar (2011). Coded SMOTE algorithm, matlab file exchange, www.mathworks.com/matlabcentral/fileexchange/38830-smote-synthetic-minority-over-sampling-technique.

Manjula, K.A. (2015). Role of image segmentation in digital image processing for information processing, *International Journal of Computer Science Trends and Technology (IJCST)*, 3(3): 312. ISSN: 2347-8578, www.ijcstjournal.org.

Manju, K., and R.S. Sabeenian (2018). Robust CDR calculation for glaucoma identification, *Biomedical Research*, Special Issue: S137–S144, ISSN 0970-938X.

Maya Alsheh, Ali, Thomas Hurtut, Timothée Faucon, and Farida Cheriet (2014). Glaucoma detection based on local binary patterns in fundus photographs, *SPIE Medical Imaging*, February 2014, United States, pp. 903531-903531-7, 2014. doi:10.1117/12.2043098. <hal-00993552>.

Matlab background, http://cimss.ssec.wisc.edu/wxwise/class/aos340/spr00/whatismatlab.htm, visited at (January, 2018).

Medha, V., and M. Pradeep (2014). Performance evaluation of optic disc segmentation algorithmsin retinal fundus images: An empirical investigation, *International Journal of Advanced Science and Technology*, 69: 19–32.

Morris, T., and Suraya Mohammed (2015). Characterizing glaucoma using texture, the University of Manchester, pp.1–6.

Morton, M.J., D.H. Whaley, and K.R. Brandt (2002). The effects of computer aided detection (CAD) on a local/regional screening mammography program: Prospective evaluation of 12,646 patients. *Radiology*, 225(P): 459.

Muramatsu, C., Y. Hatanaka, A. Sawada, T. Yamamoto and H. Fujita (2011). Computerized detection of peripapillary chorioretinal atrophy by texture analysis, 33rd Annual International Conference of the IEEE EMBS, Boston, pp. 5974–5950.

Chanel Murugan, Bomikazi Z. Golodza, Kaveshni Pillay, Brightness N. Mthembu, Praneal Singh and Sibusiso K. Maseko (2015). Retinal nerve fibre layer thickness of black and Indian myopic students at the University of KwaZulu-Natal, *African Vision and Eye Health*, 74(1): 6, Art. #24.

Muthukrishnan, R., and M. Radha (2011). Edge detection techniques for image segmentation, *International Journal of Computer Science & Information Technology*, 3(6): 259.

Murugan, C., B.Z. Golodza, K. Pillay, B.N. Mthembu, P. Singh, S.K. Maseko, S. Jhetam, and N. Rampersad (2015). Retinal nerve fiber layer thickness of black and Indian myopic students at the University of KwaZulu- Natal. *Vision Eye Health*, 74(1), Art. #24, 6 pages. doi:10.4102/aveh.v74i1.24

Nawaldgi, Sharanagouda, and Y.S. Lalitha (2017). A novel combined color channel and ISNT rule based automatic glaucoma detection from color fundus images, *Indian Journal of Science and Technology*, 10(13). doi:10.17485/ijst/2017/v10i13/111722.

Nasiruddin, Mohd, Faizan Ahmed, Ashar Quazi, Mubashara Mehrosh, and Kahekashan Anjum (2017). Computer aided design of glaucoma detection, *International Journal of Engineering Science and Computing*, 7(3).

Naz, Sobia, and Sheela N. Rao (2014). Glaucoma detection in color fundus images using cup to disc ratio, *The International Journal of Engineering and Science (IJES)*, 3(6): 51–58. ISSN (e): 2319-1813 ISSN (p): 2319–1805.

Ndajah, Peter, Hisakazu Kikuchi, Masahiro Yukawa, Hidenori Watanabe, and Shogo Muramatsu (2011). An investigation on the quality of denoised images, *International Journal Of Circuits*, 5(4): 423–434.

Nicholas, M.J., P.J. Slanetz, and J.B. Mendel (2004). Prospective assessment of computer-aided detection in interpretation of screening mammography, *AJR*, 182(P): 32–33.

Nouri-Mahdavi, K., D. Hoffman, D.P. Tannenbaum, S.K. Law, and J. Caprioli (2004). Identifying early glaucoma with optical coherence tomography, *American Journal of Ophthalmology*, 137(2): 228–235.

Odstrcilik, J.,R. Kolar,V. Harabis, J. Gazarek, and J. Jan (2010). Retinal nerve fiber layer analysis via markov random fields texture modelling, Proceedings of the 18th European Signal Processing Conference, pp. 1650–1654.

Odstrcilika, Jan, Radim Kolara, Ralf-Peter Tornowc, *et al.* (2014). Thickness related textural properties of retinal nerve fiber layer incolor fundus images, *Computerized Medical Imaging and Graphics*, 38(6): 508–516.

Oh, J.E., H.K. Yang, K.G. Kim, and J.-M. Hwang (2015). Automatic computer-aided diagnosis of retinal nerve fiber layer defects using fundus photographs in optic neuropathy. *Investigative Ophthalmology & Visual Science*, 56(5): p.2872–2879.

Otsu, N. (1979). A threshold selection method from gray-level histograms, *IEEE Transactions on Systems, Man, and Cybernetics*, 9(1): 62–66.

Pachiyappan, Arulmozhivarman, Undurti N. Das, Tatavarti V.S.P. Murthy, and Rao Tatavarti (2012). Automated diagnosis of diabetic retinopathy and glaucoma using fundus and OCT images, *Lipids in Health and Disease*, 11: 73.

Pallawala, P., W. Hsu, M. Lee, and K. Eong (2004). Automated optic disc localization and contour detection using ellipse fitting and wavelet transform, Proceedings of European Conference on Computer Vision, pp. 139–151.

Paranjape, Raman B. (2009). *Handbook of Medical Image Processing and Analysis*, 2nd ed.

Patil, Dnyaneshwari D., Ramesh Manza, and Gangadevi C. Bedke (2014). Diagnose glaucoma by proposed image processing methods, *International Journal of Computer Applications* (0975-8887) 106(8), pp. 14–17.

Poon, Linda Yi-Chieh, David Solá-Del Valle, Angela V. Turalba, Iryna A. Falkenstein, Michael Horsley, Julie H. Kim, Brian J. Song, Hana L. Takusagawa, Kaidi Wang, and Teresa C. Chen (2017). The ISNT rule: How often does it apply to disc photographs and retinal nerve fiber layer measurements in the normal population? *American Journal of Ophthalmology*, 184: 19–27.

Prageeth, P.G., J. David, and A. Sukesh Kumar (2011). Early detection of retinal nerve fiber layer defects using fundus image processing, Proceedings of the IEEE Recent Advances in Intelligent Computational Systems (RAICS '11), (September 2011), pp. 930–936, Trivandrum, Kerala.

Prasantha, H.S., H.L. Shashidhara, K.N.B. Murthy, and Lata G. Madhavi (2010). Medical image segmentation, *(IJCSE) International Journal on Computer Science and Engineering*, 02(04): 1209–1218.

Priyadharshini, M., L. Glory, and J. Anitha (2014). A region growing method of optic disc segmentation in retinal images, International Conference on Electronics and Communication Systems, pp. 1–5.

Pudil, P., J. Novovičová, and J. Kittler (1994). Floating search methods in feature selection, *Pattern Recognition Letters*, 15(11): 1119–1125.

Quellec, G., K. Lee, M. Dolejsi, M.K. Garvin, M.D. Abràmoff, and M. Sonka (2010). Three-dimensional analysis of retinal layer texture: Identification of fluid-filled regions in SD-OCT of the macula, *IEEE Transactions on Medical Imaging*, 29(6): 1321–1330.

Quigley, H.A. (1999). Neuronal death in glaucoma, *Progress in Retinal and Eye Research*, 18(1): 39–57.

Qureshi, Imran (2015). Survey: Glaucoma detection in retinal images using image processing techniques, *The International Journal of Advanced Networking and Applications*, 7(02): 2705–2718. ISSN: 0975-0290.

Rajaiah, Preethi, and R. John Britto (2014). Optic disc boundary detection and cup segmentation for prediction of glaucoma, *International Journal of Science Engineering and Technology Research (IJSETR)*, 3(10): 2665–2671.

Ramaswamy, Akshaya, Keerthi Ram, and Mohanasankar Sivaprakasam (2016). A depth based approach to glaucoma detection using retinal fundus images. In: X. Chen, M.K. Garvin, J. Liu, E. Trucco, Y. Xu (Eds.) Proceedings of the Ophthalmic Medical Image Analysis Third International Workshop, OMIA 2016, Held in Conjunction with MICCAI 2016, Athens, Greece, October 21, pp. 9–16. doi:10.17077/omia.1041.

Ranjith, N., C. Saravanan, and M.R. Bibin (2015). Glaucoma diagnosis by optic cup to disc ratio estimation, *International Journal of Inventive Engineering and Sciences (IJIES)*, 3(5): 1–5. ISSN: 2319-9598.

Ravi, S., and A.M. Khan (2013). Morphological operations for image processing: Understanding and its applications, Vignan's University, NCVSComs-13 Conference Proceedings, pp. 17–19.

Ritch, R., and J.M. Liebermann (1999). Angle closure glaucoma, *Asian Journal of Ophthalmology*, 1(3): 10–16.

Ritch, R., J. Liebmann, and C. Tello (1995). A construct for understanding angle-closure glaucoma: The role of ultrasound bio microscopy, *Ophthalmology Clinics of North America*, 8: 281–293.

Sahoo, P.K., S.A.K.C. Soltani, and A.K. Wong (1998). A survey of thresholding techniques, *Computer Vision, Graphics and Image Processing*, 41(2): 233–260.

Sahu, A., G. Runger, and D. Apley (2011). Image denoising with a multi-phase kernel principal component approach and an ensemble version, IEEE Applied Imagery Pattern Recognition Workshop, pp. 1–7, Washington, DC.

Saine, Patrick J., and Marshall E. Tyler (2014). *Fundus Photography Overview*. Ophthalmic Photography Society, ISBN: 0750673729. Available at www.opsweb.org/page/fundusphotography.

Sakthivel, K., and R. Narayanan (2015). An automated detection of glaucoma using histogram features, *International Journal of Ophthalmology*, 8(1): 194–200.

Salam, A.A., T. Khalil, M.U. Akram, A. Jameel, and I. Basit (2016). Automated detection of glaucoma using structural and non- structural features, *Springer Plus*, 5(1): 1519.

Salam, Anum A., Tehmina Khalil, M. Usman Akram, Amina Jameel, and Imran Basit (2016). Automated detection of glaucoma using structural and nonstructural features, *Springerplus*, 5(1): 1519. Published online 2016 September 9. doi:10.1186/s40064-016-3175-4, PMCID: PMC5017972, PMID: 27652092.

Sample, P.A., C.F. Bosworth, E.Z. Blumenthal, C. Girkin, and R.N.Weinreb (2000). Visual function-specific perimetry for indirect comparison of different ganglion cell populations in glaucoma, *Investigative Ophthalmology & Visual Science*, 41(7): 1783–1790.

Satish, T., and J. Sunita (2015). Optic disc and cup segmentation for glaucoma screening based on superpixel classification, *International Journal of Innovations & Advancement in Computer Science*, 4(Special Issue). IJIACS ISSN 2347-8616.

Saxena, Swati, and R.L. Yadav (2015). Hybrid feature of tamura texture based image retrieval system, *International Journal of Recent Research and Review*, VIII(2). ISSN 2277-8322, pp. 23–28.

Seiffert, Chris, Taghi M. Khoshgoftaar, Jason Van Hulse, and Amri Napolitano (2010). Rusboost: A hybrid approach to alleviating class imbalance, *IEEE Transactions on Systems, Man, and Cybernetics—Part A: Systems and Humans*, 40(1), 185–197.

Senthilkumaran, T., and R. Rajesh (2009). Edge detection techniques for image segmentation–a survey of soft computing approaches, *International Journal of Recent Trends in Engineering*, 1(2), pp. 250–254.

Septiarini, Anindita, and Agus Harjoko (2015). Automatic glaucoma detection based on the type of features used: A review, *Journal of Theoretical and Applied Information Technology*, 2872, pp. 366–375.

Septiarini, Anindita, and Agus Harjoko (2015). Automatic glaucoma detection based on the type of features used: A review, *Journal of Theoretical and Applied Information Technology*, 72(3). © 2005–2015, JATIT & LLS, ISSN: 1992-8645.

Septiarini, Anindita, Dyna M. Khairina, Awang H. Kridalaksana, and Hamdani Hamdani (2018). Automatic glaucoma detection method applying a statistical approach to fundus images, *Healthcare Informatics Reasearch*, 24(1): 53–60, pISSN: 2093-3681.

Sevastopolsky, Artem (2017). Optic disc and cup segmentation methods for glaucoma detection with modification of U-Net convolutional neural network, (4 April), Lomonosov: Moscow State University, arXiv: 1704.00979v1.

Shields, M.B. (2008). Normal-tension glaucoma: Is it different from primaryopen-angle glaucoma? *Current Opinion in Ophthalmology*, 19(2): 85–88.

Shinde, Amit, Anshuman Sahu, Daniel Apley, and George Runger (2014). Pre images for variation patterns from kernel PCA and bagging, *IIE Transactions*, 46(5), pp. 429–456.

Singh, P., and R.S. Chadha (2013). A novel approach to image segmentation, *International Journal of Advanced Research in Computer Science and Software Engineering Research Paper*, 3(4).

Sivaswamy, Jayanthi, S.R. Krishnadas, Gopal Dutt Joshiy, Madhulika Jainy, Ujjwaly, and A. Syed Tabish (2011). DRISHTI-GS: Retinal image dataset for optic nerve head (ONH), Arvind Eye Hospital, Madurai, India.

Sivaswamy, Jayanthi, Subbaiah Krishnadas, Gopal Joshi, Madhulika Jain, and A. Ujjwaft Syed Tabish (2014). DRISHTI-GS: Retinal image dataset for optic nerve head(ONH) segmentation, 2014 IEEE 11th International Symposium on Biomedical Imaging, ISBI 2014, pp. 53–56. doi:10.1109/ISBI.2014.6867807.

Soille, P. (1999). *Morphological Image Analysis: Principles and Applications*. Springer, pp. 164–165.

Sommer, A., J.M. Tielsch, J. Katz, H.A. Quigley, J.D. Gottsch, J. Javitt, and K. Sing (1991). Relationship between intraocular pressure and primary open angle glaucoma among white and black Americans: The Baltimore eye survey, Archives of *Ophthalmology*, 109(8): 1090–1095.

Stricker, M., and M. Orengo (1995). Similarity of color images, SPIE Conference on Storage and Retrieval for Image and Video Databases III, Vol. 2420, (February 1995), pp. 381–392, San Jose, CA.

Suero, Angel, Diego Marin, Manuel E. Gegundez-Arias, and Jose M. Bravo (2013). Locating the optic disc in retinal images using morphological techniques, IWBBIO 2013 Proceedings, Granada, 18–20 March, pp. 593–600.

Sui, Yuan, Ying Wei, and Dazhe Zhao (2015). Computer-aided lung nodule recognition by SVM classifier based on combination of random undersampling and SMOTE, *Computational and Mathematical Methods in Medicine*, 2015: 13. Article ID 368674. Hindawi Publishing Corporation.

Sundari, B., and S. Sivaguru (2017). Early detection of glaucoma from fundus images by using MATLAB® GUI for diabetic retinopathy, *International Journal of Innovative Research in Computer and Communication Engineering*, 5(1), An ISO 3297: 2007 Certified Organization, pp. 174–181.

Syed, M., and H. Kwang (2016). Depth edge detection by image-based smoothing and morphological operations, *Journal of Computational Design and Engineering*, 3: 191–197. www.sciencedirect.com.

Swapna, P.P., and M.G. Mini (2016). A regression neural network based glaucoma detection system using texture features, *International Journal of Computing Communications and Instrumentation Engineering*, 3(2). ISSN 2349-1469 EISSN 2349-1477, pp. 276–279.

Tamura, H., and T. Yamawaki (1978). Textural features corresponding to visual perception, *IEEE Transactions on Systems, Man and Cybernetics*, 8(6): 460–473.

Tan Jen Hong, U. Rajendra Acharya, Sulatha V. Bhandary, Kuang Chua Chua, and Sobha Sivaprasad (2017). Segmentation of optic disc, fovea and retinal vasculature using a single convolutional neural network, arXiv.org > cs > arXiv:1702.00509, 1–24.

Thakkar, Kartik, Kinjan Chauhan, Anand Sudhalkar, and Ravi Gulati (2017). Detection of glaucoma from retinal fundus images by analyzing ISNT measurement and features of optic cup and blood vessels, *International Journal of Engineering Technology Science and Research (IJETSR)*, 4(7): 487–493, ISSN: 2394-3386.

Thomas, Ravi, Klaus Loibl, and Rajul Parikh (2011). Evaluation of a glaucoma patient, *Indian Journal of Ophthalmology*, 59(Suppl.): S43–S52.

Uchida, H., S. Ugurlu, and J. Caprioli (1998). Increasing peripapillary atrophy is associated with progressive glaucoma, *Ophthalmology*, 105(8): 1541–1545.

van Eijnatten, Maureen, Juha Koivisto, Kalle Karhu·Tymour Forouzanfar, and Jan Wolff (2017). The impact of manual threshold selection in medical additive manufacturing, *International Journal of Computer Assisted Radiology and Surgery*, 12: 607–615.

van Ginneken, Bram (2001). Computer-aided diagnosis in chest radiography, Ph.D. Theses Abstracts, www.isi.uu.nl/;bram/thesis.

Weinreb, R.N., and P.T. Khaw (2004). Primary open-angle glaucoma, *The Lancet*, 363(9422): 1711–1720.

Wong, D.W.K., Jiang Liu, Joo Hwee Lim, Ngan Meng Tan, Zhuo Zhang, Huiqi Li, Shijian Lu, and Tien Yin Wong (2010). Method of detecting kink-bearing vessels in a retinal fundus image (CDR), The 5th IEEE Conference on Industrial Electronics and Applications (ICIEA), (June 2010), pp. 1690–1694.

Xu, J., O. Chutatape, E. Sung, C. Zheng, and P.C.T. Kuan (2007). Optic disk feature extraction via modified deformable model technique for glaucoma analysis, *Pattern Recognition*, 40: 2063–2076.

Yang, Mingqiang, Kidiyo Kpalma, and Joseph Ronsin (2008). A survey of shape feature extraction techniques, Peng-Yeng Yin. Pattern Recognition, IN-TECH, pp. 43–90, hal-00446037.

Yazdanpanah, A., G. Hamarneh, B. Smith, and M. Sarunic (2009). Intra-retinal layer segmentation in optical coherence tomography using an active contour approach, *Medical Image Computing and Computer-Assisted Intervention*, 12(2): 649–656.

Ying, Huajun, Ming Zhang, and Jyh-Charn Liu (2007). Fractal-based automatic localization and segmentation of optic disc in retinal images, 28th Annual International Conference of the IEEE, Engineering in Medicine and Biology Society (EMBS).

Yogamangalam, R., and B. Karthikeyan (2013). Segmentation techniques comparison in image processing, *International Journal of Engineering and Technology (IJET)*, 5(1): 307–313.

Youssif, A.A.A., A.Z. Ghalwash, and A.A.S.A. Ghoneim (2008). Optic disc detection from normalized digital fundus images by means of a vessels' direction matched filter, *IEEE Transactions on Medical Imaging*, 27(1): 11–18.

Yucel, Y.H., Q. Zhang, R.N. Weinreb, P.L. Kaufman, and N. Gupta (2003). Effects of retinal ganglion cell loss on magno, parvo, konio cellular path ways in the lateral geniculate nucleus and visual cortex in glaucoma, *Progress in Retinal and Eye Research*, 22(4): 465–481.

Zhang, Zhuo, Liu Jiang, N.S. Cherian, Ying Sun, Joo Hwee Lim, Wing Kee Wong, Ngan Meng Tan, Shijian Lu, Huiqi Li, and Tien Ying Wong (2009). *Convex Hull Based Neuro-Retinal Optic Cup Ellipse Optimization in Glaucoma Diagnosis*. Conf Proc IEEE Eng Med Biol Soc. 2009, 1441–1444. doi:10.1109/iembs.2009.5332913. PMID: 19963748.

Zhang, Z., C.K. Khow, J. Liu, Y.L.C. Cheung, and T. Aung (2012). Automatic glaucoma diagnosis with mRMR-based feature selection, *Journal of Biometrics and Biostatistics S*, 7: 008. doi:10.4172/2155-6180.S7-008.

Zhang, Zhuo, Ruchir Srivastava, Huiying Liu, Xiangyu Chen, Lixin Duan, Damon Wing Kee Wong, Chee Keong Kwoh, Tien Yin Wong, and Jiang Liu (2014). A survey on computer aided diagnosis for ocular diseases, *BMC Medical Informatics and Decision Making*, 2–29. Copyright © www.biomedcentral.com/1472-6947/14/80.

Zhou, Zhi-Hua *Ensemble Learning*, Nanjing: National Key Laboratory for Novel Software Technology, Nanjing University, pp. 1–5. China zhouzh@nju.edu.cn.

Zhou, W., A.C. Bovik, H.R. Sheikh, and E.P. Simoncelli (2004). Image qualifty assessment: From error visibility to structural similarity, *IEEE Transactions on Image Processing*, 13(4): 600–612.

Zou, Kelly H., Simon K. Warfield, Aditya Bharatha, Clare M.C. Tempany, Michael R. Kaus, Steven J. Haker, William M. Wells III, Ferenc A. Jolesz, and Ron Kikinis (2006). Statistical validation of image segmentation quality based on a spatial overlap index, *Academic Radiology*, 11(2), 178–189. NIH Public Access Author Manuscript, available in PMC (2006 March 28).

Zuiderveld, K. (1994). Contrast limited adaptive histogram equalization. In: *Graphics Gems IV*. Academic Press, pp. 474–485.

WEBSITE

AdiBronshtein, Data Scientist. https://medium.com/@adi.bronshtein/a-quick-introduction to k-nearest-neighbors algorithm-62214cea29c7, April (January, 2017).

AucklandUniversity. www.cs.auckland.ac.nz/courses/compsci773s1c/lectures/Image Processing-html/topic4.htm.

Channel (digital image) (2018). https://en.wikipedia.org/wiki/Channel_ (digital image), last edited on (9 February), at 08:37.

Computerized Medical Imaging and Graphics, (2014). Vol. 38, pp. 508–516, Copyright © www.biomedical-engineering-online.com.

Dataman, Using Under-Sampling Techniques for Extremely Imbalanced Data. https://towardsdatascience.com/sampling-techniques-for-extremely-imbalanced-data-part-i-under-sampling-a8dbc3d8d6d8 (September, 2019).

Davies, E.R. (1997). *Machine Vision: Theory, Algorithms, Practicalities, Elsevier, 4th ed.*

Devi, H. (2006). Thresholding: A pixel-level image processing methodology preprocessing technique for an OCR System for the Brahmi Script, *Ancient Asia*, 1, pp.161–165.

Eye source. www.ammoparadise.com/anatomy-and-physiology-of-the-human-eye/ (April, 2017).

Figure 1.1 Shutterstock website. www.shutterstock.com/image-photo/medical-image-ocular-fundus-human-retina-1126908659, figure1.

Figure 2.1 Shutterstock website. www.shutterstock.com/image-vector/human-eye-anatomy-infographics-outside-view-1051429736.

Figure 2.2 Shutterstock website. www.shutterstock.com/image-vector/structure-human-eye-organization-retina-optic-1135203455, figure2.

Figure 2.3 Shutterstock website. www.shutterstock.com/image-photo/fluorangiography-left-eye-central-serous-chorioretinopathy-1256153611, figure 4.

Figure 2.5 Shutterstock website. www.shutterstock.com/image-photo/optical-coherence-tomography-oct-image-eye-431734615, figure1.

Figure 2.6a Shutterstock website. www.shutterstock.com/image-photo/ophthalmic-image-detailing-retina-optic-nerve-10136503, figure2.

Figure 2.6b Shutterstock website. www.shutterstock.com/image-photo/medical-image-retinashowing-optic-disc-cupping-190245218, figure3.

Figure 2.7 Shutterstock website. www.google.com/search?q=normal+optic+nerve+head&safe=strict&sxsrf=ACYBGNSrJfzs0hbd4XvxRG9wqnh-VxrjLQ:1577123726741&source=lnms&tbm=isch&sa=X&ved=2ahUKEwi91P7bq8zmAhVOrxoKHWfZAiMQ_AUoAXoECA4QAw&biw=1366&bih=657#imgrc=tk0Ly6Y-tNGuaM:, figure1.

Figure 2.8 Shutterstock website. https://jirehdesign.com/stock-eye-illustrations/eye-disease-illustrations/optic-nerve-disease/optic-disc-cupping-progression-co0052/?v=3dd6b9265ff1, figure1.

GLCM alogorithm. (https://support.echoview.com/Web Help/Windows_and_Dialog_Boxes/Dialog_Boxes/Variable_properties_dialog_box/Operator_pages/GLCM_Texture_Features.htm), visited at (August, 2018).

GLCM MATLAB® code. www.mathworks.com/matlabcentral/fileexchange/22187-glcm-texture-features Copyright © 2008, Avinash Uppuluri, All rights reserved.

GLCM Texture Feature. https://support.echo view.com ComebHelp/Windows_and_Dialog_Boxes/Dialog_Boxes/Variable_properties_dialog_box/Operator_pages/GLCM_Texture_Features, Wednesday 8th November 2017.

Halalli, Bhagirathi, and Aziz Makandar (2017). *Computer Aided Diagnosis – Medical Image Analysis Techniques.* http://dx.doi.org/10.5772/intechopen.69792.

Heidelberg engineering. www.google.com/search? safe=strict&client =firefox- b&biw= 1252&bih=600&tbm =isch&sa=1&ei=sQNcW_ OoHouSaMnFiPAO&q=Heidelberg+Retina+Tomography+HRT+image&oq=Heidelberg+Retina+Tomography+HRT+image&gs_l=img.12...33928.36238.0.38324.2.2.0.0.0.0.1211.1473.2-1j7-1.2.0.... 0... 1c.1.64.img..0.0.0....0.M-bc1KS CONE#imgrc= RpiyWY0y9tmOK M (December, 2017).

Human eye structure. www.shutterstock.com (April, 2017).

Math Works. www.mathworks.com/discovery/image segmentation. Html (February, 2018).

Median filter. https://en.wikipedia.org/wiki/Median_filter (November, 2017).

Morphological image processing. www.cs.auckland.ac.nz/courses/compsci773s1c/lectures/ImageProcessinghtml/topic4.htm?_ga=2.48742822.1040376964.153184491444313370 5.1524854676 (December, 2016).

Ophthalmic Photographers Society. www.opsweb.org/page/fundusimaging (August, 2019).

Practical Guide to deal with Imbalanced Classification Problems. www.analyticsvidhya.com/blog/2016/03/practical-guide-deal-imbalanced-classification-problems/ (July, 2018).

Ray, Sunil. www.analyticsvidhya.com/blog/2017/09/understaing-support-vector-machine-example-code/), visited at (October, 2017).

Regions properties function. https://octave.Source forge. Io/image/function/regionprops.html (May, 2017).

Retina Image Bank. http://imagebank.asrs.org/discover-new/ files/4/ 25? q = fundus %20auto-fluorescence%20(faf) (May, 2018).

Retinal imaging device modified from smart phone. www.google.com/search?q=fundus+camera+using+smart+phone&safe=strict&client=firefox-b&source=lnms&tbm=isch&sa=X&ved=0ahUKEwiV4uzgisHcAhUPUBoKHS47DpsQ_AUICigB&biw=1252&bih=600#imgrc=gwi-W6w9uaApzM (July, 2018).

Roerdink, J.B.T.M., and A. Meijster (2000). The watershed transform: Definitions, algorithms and parallelization strategies, *Fundamenta Informaticae*, 41: 187–228.

Sidana, Mandy, Types of classification algorithms in Machine Learning. https://medium.com/@Mandysidana/machine-learning-types-of-classification-9497bd4f2e14 (September, 2019).

Tamura Features MATLAB®, Coded by Sudhir Sornapudi. https://github.com/Sdhir/TamuraFeaturesm, Copyright © 2014, Sudhir Sornapudi, All rights reserved.

The MathWorks, Inc. Marker-Controlled Watershed Segmentation, [Online]. Available: www.m athworks.com/ products/demos/ image/watershed/ipexwatershed.html (August, 2013).

What are the mean and median filters. www.markschulze.Net/java/meanmed. html (November, 2017).

Index

For Product Safety Concerns and Information please contact our EU
representative GPSR@taylorandfrancis.com Taylor & Francis Verlag GmbH,
Kaufingerstraße 24, 80331 München, Germany

Printed and bound by CPI Group (UK) Ltd, Croydon, CR0 4YY
01/05/2025
01858522-0001

018584040-0001

01/05/2025

Printed and bound by CPI Group (UK) Ltd, Croydon, CR0 4YY

For Product Safety Concerns and information please contact our EU
representative GPSR@taylorandfrancis.com Taylor & Francis Verlag GmbH,
Kaufingerstraße 24, 80331 München, Germany